SPECTRUM POLITICS

SPECTRUM POLITICS
Unveiling the Defence

SALMAN KHURSHID
Daksha Sharma

RUPA

Published by
Rupa Publications India Pvt. Ltd 2018
7/16, Ansari Road, Daryaganj
New Delhi 110002

Sales centres:
Allahabad Bengaluru Chennai
Hyderabad Jaipur Kathmandu
Kolkata Mumbai

Copyright © Salman Khurshid and Daksha Sharma 2018

The views and opinions expressed in this book are the authors' own and the facts are as reported by them which have been verified to the extent possible, and the publishers are not in any way liable for the same.

All rights reserved.
No part of this publication may be reproduced, transmitted, or stored in a retrieval system, in any form or by any means, electronic, mechanical, photocopying, recording or otherwise, without the prior permission of the publisher.

ISBN: 978-93-5304-050-5

First impression 2018

10 9 8 7 6 5 4 3 2 1

The moral right of the authors have been asserted.

Printed by Parksons Graphics Pvt. Ltd., Mumbai

This book is sold subject to the condition that it shall not, by way of trade or otherwise, be lent, resold, hired out, or otherwise circulated, without the publisher's prior consent, in any form of binding or cover other than that in which it is published.

To,
Sam Pitroda, who changed the landscape and put airwaves into battle for development.

CONTENTS

Preface	ix
1. Of Missed Calls and Wrong Numbers	1
2. Decoding Spectrum Policy	11
3. 2G Case: The Bare Facts	36
4. Facts from Justice Shivraj V. Patil's Report	65
5. Colours of the Rainbow: The JPC Report	89
6. Dark Clouds Over Sunshine Sector	110
7. Presidential Reference	134
8. The Final Vindication	145
Epilogue: Curtain Call	159
Acknowledgements	165
Abbreviations	167
Index	171

PREFACE

We live in times of great irony. The staggering democratic decision in the 2014 General Election seemingly changed the course of history, sweeping away many beliefs considered to be immutable dimensions of the idea of India. The winning combination quickly assumed it to be, at long last, recognition of India's true character and repudiation of what they believed was adulteration of our national ethos. The extent and impact of the change left large numbers numbed into silence, and even the hardy campaigners of liberal India retreated to reconsider strategies and regrouped for resisting the unthinkable that had happened.

Unfortunately, the waves of adversity kept advancing as state after state fell to the 'Modi magic.' It was only the spirited effort of Congress President Rahul Gandhi in Gujarat—an election throwing up a no-holds-barred combat between the 'Idea of India' versus some ambivalent notion of 'Asli Bharat'—that put up a real challenge to the Bharatiya Janata Party (BJP) within its traditional stronghold. Our failure by a whisker to snatch Gujarat from the BJP after three terms in office, showed the path towards success in 2019 flagging the weaknesses we need to overcome by then. There is an organizational imperative, the need to garner finances for politics that is rapidly becoming more expensive and perhaps the magic formula for a suitable alliance.

For people who cannot or will not desert the ship even in desperate times, it is important to fortify hope without which neither their resolve nor the vessel they resolutely defend would survive. For that, it is important to disbelieve false narratives about the ship and its crew, expose misrepresentation, retrieve truth and find the right idiom to convey it to the nation. Since public discourse has, of late, become prickly and divisive with much of the electronic media taken to supporting reductionist logic in public affairs, platforms for the alternative points of view are rapidly shrinking. Therefore, a tactical mix of old and new media instruments shall have to be garnered.

In contemporary India, diversity is under attack by some associated with the current establishment and left undefended by others for the fear of further defeats. The much-touted ambition of the BJP to achieve a 'Congress mukt Bharat' by wiping out all traces of what India had achieved under successive Congress governments is most certainly a delusion, but the threat to the idea of India born from our Independence movement is quite real, if events of recent months are anything to go by.

It is, therefore, imperative for the patriotic citizens of India to wage a battle to preserve diversity, which is critical to our understanding of our Constitution. In this enterprise, the citizens have to come equipped with the ability to tell the right from the wrong, show conviction for the right thing to do and stand up and be counted for people's rights. To keep our perspectives clear, we will need to cut through reams of misinformation and fake news, not to mention post-truth. Unity is India's strength, but enforced uniformity sought to be pursued by the BJP and RSS may well be her bane, particularly if based on false notions of justice.

The fight for justice is as much in courts of law as in the courts of public opinion. The highest court, they say, is the court of the people. Arguments before a judge are quite different from those before citizens; the latter are not directly relevant before courts of law, although the former can often be helpful in public debates. I have attempted to put the entire 2G spectrum case before the people, judgements and all, in a manner that makes it comprehensible, and thereafter, to rest the case.

I have highlighted the decisions of the BJP-led National Democratic Alliance (NDA) regime from 1998 to 2004 in the telecom sector and have underscored their continuity under the United Progressive Alliance (UPA) that the BJP conveniently obfuscates. The 2G case was far from a scandal; it was rather a corporate battle to scuttle the allocation of spectrum licences on the recommendation of the Telecom Regulatory Authority of India (TRAI)[1]. The central bone of contention in this case was the manner in which spectrum licences were allocated to the applicants, which resulted in wild, baseless, unsubstantiated as well as

---

[1]TRAI is a body corporate consisting of a chairperson and not more than two part-time members appointed by the Central Government.

unjustified allegations and accusations of cheating, forgery, corruption and criminal conspiracy. It must be highlighted that those castigating the Congress-led UPA regime are surely themselves aware of the origin of the discriminatory policy and also the fact that the UPA government merely followed the telecom policy established by its predecessors. The decisions taken by the preceding government clearly establish that the trend of not auctioning spectrum was already entrenched and was consistently followed under the UPA regime.

In this regard, what needs to be pertinently analysed is the proposition that whether auctioning the 2G spectrum and distributing it on a first-come-first-serve basis is to be considered an erroneous act, not to speak of corruption, given that government policy was greater penetration rather than earning revenue from an essential infrastructure for development. If the answer is in the negative, then there is no further reason, rationality and legality that can be attributed to the arguments of corruption in this so-called scam.

In the light of the circumstances objectively explained and in the absence of any material evidence of wrongdoing noted by the trial court, readers shall appreciate that the 2G spectrum issue was blown out of proportion by a few political parties as well as a section of the media. They compromised national interests to derive political mileage from this controversy.

This book is also an attempt to remove the acute confusion being created in the minds of the people of India— the ultimate creators of our nation's destiny. Contrary to public perception, the story of the telephony revolution did not start with the Telecom Ministry's decision of 2007–08 unfairly described as a scam. In fact, it started in 1994 when the first mobile telephony licences were granted and India set out to capture a place in the communications revolution. The seeds of this revolution were planted way back during the leadership of Rajiv Gandhi who looked far and wide to pick the best brains like Sam Pitroda as the iconic Head of C-Dot who subsequently became advisor to Rajiv Gandhi and was responsible for shaping India's foreign and domestic telecommunications policies. Similarly, in many ways, the negative outcomes that rattled the UPA government because of 2G and coal allocations ironically have their roots in the economic reforms of the Congress government under Prime Minister P.V. Narasimha Rao and were taken further under UPA-I. Divesting the State of its direct control of the economy was targeted at unleashing the potential of growth in the economy that had artificially been kept in check in order to control social disparity. Undoubtedly, in the present endeavour, every effort has been made to establish the true facts of the 2G case so that readers become aware of how deliberate distortion caused endless anguish to individuals, the Congress party and eventually the country.

The contents of this book, therefore, are in a sense about the past, present and also about the future. Despite being advised by several well-meaning persons to steer clear of the hostile terrain of public perception, I chose to plunge into troubled waters, if for nothing else but to place on record that we, who were part of the UPA, too have something to say. As our then Prime Minister Dr

Manmohan Singh used to say, ultimately history will judge us for what we intended, achieved and yet, for which, we were unfairly castigated.

1
OF MISSED CALLS AND WRONG NUMBERS

Although the entire period of five years of UPA-II were dubbed as wasted years by the opponents who rode to victory in 2014, and also by many disappointed well-wishers, there is much that can and must be salvaged. It becomes important to recover the truth when hollow and exaggerated claims in the name of development provide an alibi for shortcuts to power and its retention, even as the consensual version of social concord in society is being ripped apart cynically, perhaps even viciously.

Furthermore, for the sake of the integrity of our country's institutional history, it is necessary to mount a salvage operation to put those years in the right perspective. We owe this to succeeding generations. But that of course will require a serious look at the decisions that were taken, most of them during UPA-I, seen by many, including the electorate at that time, as a successful tenure. Unlike the

India Shining claim that could not get Prime Minister Atal Bihari Vajpayee re-elected in 2004, UPA-I was rewarded with a greater majority in the 2009 elections.

By any yardstick, those were boom years (2004–09) and should normally have been the launch pad for consolidation and greater success beyond 2009. Yet, tragically, the opposite happened. Virtually, five years of relentless and unprincipled political bludgeoning culminated in the ignominy of 2014. 'Good economics makes bad politics' became the dirge of UPA-II and with it the stifling of the most ambitious social welfare experiment that included National Rural Health Mission (NRHM), Mahatma Gandhi National Rural Employment Guarantee Act (MGNREGA), Right to Education (RTE), AADHAAR-linked Direct Benefits, et al. The triumph of the Indian Camelot ended in tragedy.

The salvage exercise demands considerable efforts since several sectors and government decisions need revisiting. But since the 2G episode was proclaimed to be the mother of all scams, we chose to put it through forensic scrutiny. As this book reached its last mile, the country was stunned by the verdict pronounced by Judge O.P. Saini acquitting all accused in the 2G trial. Predictably, there were celebrations in some quarters, while others put up a brave face but with weak logic that 'the acquittal does not mean there was no scam'. An appeal has been filed in the Delhi High Court, but given that it will be under consideration for a long time (all 1,552 pages), the political sting has been decisively blunted.

The voluminous judgement merits separate consideration, but at this stage, it must be noted that in addition to the clear acquittal by the CBI court, the two reports—first of

Justice Shivraj Patil and the second of the Joint Parliamentary Committee (JPC)—and the Five Judges Constitutional Bench Advisory Opinion to the President of India, all variously rejected the Comptroller and Auditor General (CAG) report. This report, that was relied upon by the Supreme Court Bench headed by Justice G.S. Singhvi, scrapped the 122 licences awarded under the former Telecom Minister A. Raja.

Despite such preponderance of material to vindicate our position, one questionable CAG report derailed and, in the process, may have cost India several decades of phenomenal growth that it was poised to achieve. Moreover, it might just have given an opportunity to what marginal forces were, till recent years, to radicalize in India and diminish it in unthinkable ways.

The Spark That Set Off a Devastating Fire

The rumblings of trouble began in November 2010 with the leaked CAG report on the 2G Spectrum which suggested that ₹1.76 lakh crore were lost (described as presumptive loss) to the government exchequer because of myopic, even legally questionable policies of the government in handing out valuable spectrum resource in 2007–08 at the 2001 price.

First, newspapers and news channels took the bit and ran, only to be overtaken by the Parliament that comprised an Opposition desperate to capture power and deeply divided treasury Benches. The former had, for over a decade, failed to whip up religious sentiment beyond a point, while the latter was caught in maladjusted attitudes towards the duality of power between the party and the government, something that had appeared to be a master stroke when UPA-I trounced the BJP in 2004.

Session after session, week after week, Parliament was paralysed by an unrelenting and abrasive Opposition with chants of scam and demands for the resignation of important members of the government. The party and its allies that had returned to power with an increased majority and had outstanding and ambitious plans of inclusive development, were suddenly thrown into confusion and self-doubt followed by induced inaction or what was maliciously described as policy paralysis. While on the one hand the government was not allowed to work, it was, on the other, accused of being unable to do any work. Our economic policies were obstructed, while our social policies were distorted to cause acute self-consciousness. It is ironic in the extreme that many policies that the NDA then opposed such as Aadhaar, Foreign Direct Investment (FDI) in Retail, Goods and Services Tax (GST), etc. are now being implemented (although with flaws) and being claimed as great achievements. Again, the unwholesome trend of disrupting parliamentary proceedings is haunting them despite their majority.

Admittedly, by the time the 2G affair rocked our boat, we had already been pulverized by the CWG controversy which was like handing to our opponents our destiny on a platter of meaningless and compromised integrity and incompetence. Looking back at the time of the opening ceremony, one wonders if the watchful citizen recalls the contrast of the applause the then Chief Minister Sheila Dikshit received and the sound of disapproval that greeted Suresh Kalmadi. Clearly, Dikshit with her formidable reputation for good work, retained some hold on the public imagination, but the talented Chairman of the Indian Olympic Association (IOA) had obviously slipped beyond repair. It was sad that a remarkable career that started with a commission in the Air Force and blossomed into sturdy political stature as the invincible MP of Pune and right-hand man to Sharad Pawar, came crashing down where young athletes jousted for gold and glory.

Kalmadi, whose political star rose to take him to ministership in the railways, has not recovered from the sliding slope even after a decade. On top of it came the botched attempt to deal with the slippery adversary who had mastered the art of manipulating public perception—the current Chief Minister of Delhi, Arvind Kejriwal. Ably assisted by Anna Hazare, who came, fasted and conquered, the former IRS civil servant converted the long-debated Lokpal into a battle cry for anarchy, but virtually destroyed the ability of the Congress party, itself born as a mass movement, to deal with crowds possessed by the cause of eradicating corruption from our daily lives. The deluge of cynicism and false salvation combined in a fatal mix to sweep away Dikshit despite a remarkable record of growth of services in Delhi.

But then the mood in the country had rapidly turned anti-establishment and it took a little spark to set off a fire of devastating proportions. The impassioned crowds that took over the streets of Delhi and Ramlila Maidan with the flag of India Against Corruption (IAC) were matched only by the spontaneous, leaderless young people who, in bitter winter, for days on end, thronged the central vista from India Gate to Rashtrapati Bhawan in solidarity with Nirbhaya, brutally assaulted and fatally injured by a group of maladjusted youth. Such was the dominance of the youth across Lutyens' Delhi that President Putin's formal banquet had to be shifted from Andhra Bhavan to the PM's Residence.

Suddenly, it seemed that we had lost the ability to speak while people in the streets accused us of being blind, deaf and insensitive. For days on end, Kejriwal held the national capital to ransom; fasts, flags, and feverish slogans of change injected idealism bordering on revolutionary zeal. Kejriwal became the oracle who wielded a brush (later to be replaced by the broom symbol of his newly established Aam Admi Party [AAP]) that relentlessly and mercilessly tarred many a reputation, starting with the unimpeachable and noble Prime Minister, Dr Manmohan Singh.

Huge posters with more than a dozen ministers portrayed as villains went up in the city with coarse invitations to citizens to vent their ire with invectives and shoes. Kejriwal picked out, one by one, each minister who could match his histrionics with sound logic. I myself came in for special attention for daring to answer back in the media where he was a star. He picked on a misguided, motivated report on Disabilities NGO with which I am associated, and actually travelled to my constituency, Farrukhabad, ostensibly to hurt me politically. A subsequent regret by the news channel concerned did little to prick Kejriwal's conscience.

I was intrigued that our party having had a rich heritage of mobilizing popular public support going back to the Independence movement suddenly became aliens to the streets. Long before Occupy Wall Street challenged the established order in western societies, we saw that strategy on the streets of Delhi. I recall seeking answers from the likes of Mahabal Mishra, Congress MP from North Delhi, famous for his ability to gather humongous crowds as well as Satpal Maharaj, the then Congress MP from Tehri and a spiritual guru with a vast number of devotees. Both leaders lamented that no one asked them to help as though we had no plans to react to the Kejriwal onslaught. The latter even ventured to suggest that he would have let loose an army of volunteers who would launch an unbearable assault of nature calls and drive the revolutionaries away.

Clearly, we had turned into an impervious bureaucratic machine far removed from the political moorings that made the Congress party the dominant force of Indian democracy. What makes the irony more piquant is the way in which the then Minister for Home Affairs and one of our best ministerial assets, P. Chidambaram allowed Arvind Kejriwal to outsmart him. Unlike many people who concede his great intelligence and knowledge, but consider him politically naive, I hold his political gumption in high esteem. But then we were yet to see Kejriwal as a slippery and crafty interlocutor when the Group of Ministers (GoM) was mandated to negotiate the resolution of the Lok Pal stalemate.

Although the perception battle was sadly lost well before the 2014 elections, and we went into the campaign as a tired and worn-out party, the adverse poll results of considerable magnitude did not put a closure to several issues, including spectrum allocation. The campaign to displace the UPA got over, but the campaign to malign its leaders continues, helped by the ongoing criminal proceedings that have dragged on for months and have many interesting side tales to note. Besides the criminal proceedings in the 2G and other matters each week, fresh reports of investigations against prominent Congress leaders, including some still holding important offices, continues to darken the mood of the party cadres. This Goebbels-like repetition of false and exaggerated allegations has been able to damage morale, only to be pushed further by a series of electoral setbacks.

Since democracy does not countenance a vacuum, our moral and political retreat was followed by the ascendency of a quickly assembled political outfit—AAP in Delhi and the BJP-led NDA at the Centre. It is a moot question that if both of them were working for a similar if not the same objective—the annihilation of the Congress—how is it that there is such bitter antagonism between them? However, many seasoned political commentators continue to believe that the RSS provided muscle to Kejriwal, a view fortified by the subsequent BJP governmental positions going to some dramatis personae of the movement, like Minister of State for External Affairs, V.K. Singh and Lieutenant Governor of Puducherry, Kiran Bedi.

At least at one level the distinction is clear: The BJP, many serious political observers believe, has an ideology that purports to dismantle the secular state envisioned by the Congress leadership of the Independence movement, even if imperfectly implemented because of practical constraints rather than a weak commitment; the AAP, on the other hand, has a malleable ideology that combines extreme opposites and can best be described as unmitigated opportunism.

Yet both rely heavily on the willing credulousness of the electorate suffering from ennui, impatience, desire for change, and social envy. Curiously, similar developments have been seen in other parts of the world, beginning with Brexit in the UK and the victory of President Donald Trump as determined by votes from the electoral college, if not by popular vote, in the US.

Beyond the Decisive Mandate

The 2014 mandate was a peculiar mix of aspiration, desperation, fatigue with stagnation, false projection of secularism, misdirected nationalism et al. It certainly was not about love-jihad and Romeo squads, Aadhaar cards for bovines and lynch mobs for cattle traders, having no burqas or not having girlfriends either, or even replacement of history with myths, speech control, undeclared emergency, media's willing suspension of disbelief, among other things. Yet the distant glimmer of hope such as in Bihar was soon shrouded in the gloom of Uttar Pradesh and Delhi. The theatre of demonetization was overwhelmed by the Stockholm syndrome; the hasty and heartless implementation of GST sought to be legitimized by evoking the spirit of sacrifice for the nation.

The important point is that many things that are happening today were not the mandate that PM Modi received in 2014. Ironically, his government's stated position is a commitment to 'sab ka sath, sab ka vikas' or equal opportunity for all. One wonders whether despite divisive politics of polarization of communities, the BJP feels that there is need for lip service to the India of yesteryears. In other words, BJP needs a bit of Congress to survive as indeed some people in the Congress might have begun to feel that the Congress needs a bit of the BJP to revive its fortunes.

When the ocean is choppy, one way is to duck the high waves and wait it out. Politics has similar tactical withdrawals to let time change the landscape for a comeback. But the conditions we operate under in our country may not allow that luxury. Like-minded liberal parties (I use that description advisedly in place of secular) have their jobs cut out, but they cannot hope to make an impact by keeping to the trodden path. Instead we need innovative and courageous steps, some already seen in Bihar and Uttar Pradesh.

Yet even as we reach out to a challenging future, the cobwebs of the past also need to be cleaned out. To the extent we have been wronged, we must forcefully defend our honour; if somewhere we were wrong we must honestly and courageously accept the mistake and move on.

2
DECODING SPECTRUM POLICY

Mobile telephony, an innovation of the late twentieth century, has had a dramatic impact on the lives of people. In India, mobile phones have swiftly overtaken landline instruments and changed the way we conduct our business and social life. While on the one hand it has opened up enormous opportunities of connectivity, on the other, it has caused issues of social etiquette. But its most important aspect is the democratic outreach it permits and facilitates. Access to information and services, just as State benefits, have become incredibly convenient. Therefore, like other infrastructure, mobile networks are essential to the scenario in terms of growth.

Mobile networks have developed exponentially from providing simple voice communication to supporting an array of data services, which includes SMS, emails, gaming, General Packet Radio Service (GPRS) and different mobile applications. It is interesting to note that telephony technology advancements have paved the way for communication without words, i.e. sending a message through images without the need of text.

Manifestly, mobile phones have made life more connected, easy to monitor and effortlessly productive to a large extent. Recent trends suggest that most of the communication on mobile phones happens via the Internet. It is, therefore, useful to have a comprehensive overview of the working of mobile networks across the globe.

Radio Spectrum—A Critical National Asset

Mobile networks use radio waves to communicate, transporting digitized voice or data in the form of oscillating electric and magnetic fields, called the electromagnetic field (EMF). The range of different electromagnetic waves varies according to frequency and extends from those at low frequencies such as 10 kHz up to high frequencies such as 100 GHz. Radio spectrum, a part of the atmosphere, transmits electromagnetic waves which enable transmission of all types of wireless signals. Spectrum thus relates to the radio frequencies allocated to the mobile industry and other sectors for communication over the airwaves.

In a very short time, the mobile industry has displayed tremendous potential to generate economic value and social benefit. Understandably, operators constantly urge national regulators to release sufficient and affordable spectrum in a timely manner for mobile companies so that more people can be connected at faster speeds. As spectrum is a finite, natural resource and a critical sovereign asset vital for communication, the government manages and controls its usage (typically by a regulator) by allocating bands for specific transmission purposes. In India, a great deal of spectrum is reserved for the defence sector. In this way, some part of it is allocated to commercial mobile communications, which in turn is allocated to mobile carriers based on certain criteria. Thus, the use of the airwaves in each country is controlled and managed by the government or the designated national regulatory authority which provides for the management of the radio spectrum and issues spectrum licences.

Spectrum planning for mobile services is governed by established tenets and dimensions. Governments work collectively through the International Telecommunication Union—a United Nations agency—to allocate specific bands to certain services on a global or regional basis, which are responsible for assigning future spectrum bands for mobile under the international treaty regime.[1] This helps to limit international interference as well as reduce the cost of mobile phones because it encourages nations to adopt compatible approaches that drive economies of scale. At the broadest level, spectrum is regulated in two ways, i.e. it is either managed through a spectrum licence or in some cases, it is unlicenced.

---

[1] International Telecommunication Union (ITU) allocates spectrum frequencies at the World Radiocommunication Conferences for the use of various countries. Allocations are made on a regional basis for different types of services. For the purpose of spectrum allocation, each member country submits its proposals to ITU, based on their requirements and priorities for opening of the bands. During the conference, all the proposals are discussed and decisions are taken for opening of the bands for new services or extension of the existing bands. See http://www.itu.int/en/about/Pages/default.aspx

The vast majority of radio spectrum is licenced encompassing a range of technologies that operate with enough power to allow the services to cover a relatively wide area. Giving an entity the exclusive rights to use a certain band signifies assurance of a certain quality of service on that specific frequency. Such usage rights are protected and the other entities causing interruption in the licenced area can be legally compelled to stop.

National regulators control access to this spectrum through an appropriate licencing framework. This allows an entity to be granted exclusive rights to use a certain frequency band in certain areas and at certain times. Licence holders include commercial organizations such as TV and radio broadcasters or mobile operators, whereas non-commercial organizations comprise emergency services and the military. Unlicenced frequency bands such as walkie talkies on the other hand, need no licence from the regulator as they cover short distances only and have more limited applications designated for certain specific types of use. The most notable examples of 'Unlicenced' technologies are Wi-Fi and Bluetooth, which both operate in the 2.4 GHz band. There are several others which are used for cordless telephones, baby monitors, car key fobs and garage door openers.

Allocation of Spectrum

The national regulator acting on behalf of the government is instrumental for the process of spectrum assignment. Administrative approach suggests that the regulator has overall control over choosing the assignment of spectrum through various methods like beauty contest, first-come-first-serve (FCFS) basis, lottery method, minimum reserve price, auctions and so on.

Of these methods, auction is the most common method employed by the regulators to assign a licence to specific users (e.g. a mobile operator) allowing them to use a specific frequency band in a certain area, at certain times for a specific period. There are numerous auction methods viz. ascending clock auction, combinatorial clock auction, Dutch auction, English or Japanese auction, sealed bid auction and simultaneous multi-round auction.

However, it must be noted that auction is not always considered the best method in case of spectrum allocation. On the general principle, the Advisory Opinion of the Constitutional Bench of the Supreme Court has underscored this point. At times, the other methods do take an edge over the traditional method of auctioning the spectrum depending on the geographical location, demand and supply conditions at the time, congestion and other relevant conditions.[2]

One of the moot points for spectrum management through the process of auction is that it offloads the cost burden on the end consumers, i.e. the high prices paid in

---

[2]The current system of auctioning exclusive licences may be the best way to allocate new frequencies for today but spectrum auctions may soon become technologically obsolete, economically inefficient and legally unconstitutional. An alternative is to step beyond the current paradigm of licenced exclusivity to a system of full openness of entry. See Eli Noam, 'Beyond Spectrum Auctions: Taking the next step to open spectrum access', *Telecommunication Policy*, vol. 21, no. 5, June 1997, pp. 461–75.

the auctions would eventually be loaded on to the customers. Further, in this method, service quality may be affected as the government cannot correctly estimate as to who will provide the services effectively. Undisputedly, there have been arguments suggesting that in case of auctions, only big players are benefited and natural resources become the monopoly of some leading corporate houses.

Studies conducted over a period of time reveal that in case of allotment of spectrum, fair allocation and not revenue maximization should be the primary consideration for auctions, thus ensuring the efficient, rational and optimum utilization of the scarce natural resource.[3] Fair and transparent allocation of spectrum at a reasonable cost to industry will maximize the value generated by a spectrum band, and this, in turn, has a positive impact on social as well as economic development of the nation.

Second-generation cellular technology (2G) was commercially launched on the GSM standard in Finland by Radiolinja (now part of Elisa Oyj) in 1991. 2G networks had three primary benefits over their predecessors—phone conversations were digitally encrypted; 2G systems were significantly more efficient on the spectrum usage allowing for far greater wireless penetration levels; and 2G introduced data services for mobile, starting with SMS text messages and

---

[3] See Eli Noam, 'Spectrum Auctions: Yesterday's Heresy, Today's Orthodoxy, Tomorrow's Anachronism. Taking the Next Step to Open Spectrum Access', *The Journal of Law and Economics*, vol. 41, No. S2, 1998, pp. 765–90. Also see Dan Steinbock and Eli M. Noam (eds.), *Competition for the Mobile Internet*, Springer Science & Business Media, 2011.

improved technology enabled various networks to provide additional services such as text messages, picture messages and MMS (multimedia messages). All text messages sent over 2G are digitally encrypted, allowing for the transfer of data in such a way that only the intended receiver can receive and read it.

After 2G was launched, the previous mobile wireless network systems were retroactively dubbed 1G. While radio signals on 1G networks are analog, radio signals on 2G networks are digital. Both systems use digital signalling to connect the radio towers (which listen to the devices) to the rest of the mobile system.

India's Tryst with Spectrum Auction

In the first auction for mobile licences in 1994, the DoT Government of India, issued eight Cellular Mobile Telephone Services (CMTS) licences—two in each of the four metro cities of Delhi, Mumbai, Kolkata and Chennai—for a period of 10 years. The government divided the country into 22 telecom circles with 281 zonal licences. The licensees were selected based on rankings achieved by them on the technical and financial evaluation created on parameters set out in the tender document. The DoT fixed several prerequisites for potential bidders to be eligible for the auction. The criteria included financial resources, reliability and investment in research as well as specific details such as rate of network rollout, pricing, quality and competitiveness.[4]

---

[4]https://indiankanoon.org/doc/37692759/

The licensees were required to pay a fixed licence fee for the initial three years and subsequently on the basis of the number of subscribers, subject to the minimum commitment mentioned in the tender document and licence agreement.

The licences and spectrum for the remaining 19 telecom circles was auctioned and allocated in 1995. In these circles, based on the experience of the industry and noting that Indian bidders would lack the financial muscle necessary, the government required that all potential bidders must have a foreign partner in order to be eligible. The government proceeded on the assumption that no Indian company, at that time, had the financial resources and technical knowledge to provide large scale mobile services.

Two blocks of 4.4 MHz from the 900 MHz band for GSM-based mobile services were auctioned in the 19 non-metro circles. However, the process exposed unforeseen problems with the design and rules of the auction. For instance, Koshika Telecom (operating under the brand name Usha Fone) was awarded multiple licences, even though the licence fee was $15 billion while Koshika only had an annual turnover of $60 million. Concerns were also raised about the possibility of a monopoly if a single company secured multiple licences. In order to address this concern, the auction rules were altered to prohibit a single company from operating in more than three circles. The auction for the 900 MHz band was then held again under the new rules. One can see that the learning curve of the sector was quite steep and the government had to constantly update its policies.

The licences issued mentioned that a cumulative maximum of up to 4.5 MHz in the 900 MHz bands would be permitted based on appropriate justification. There was no separate upfront charge for the allocation of spectrum to the licensees, who only paid annual spectrum usage charges subject to revision from time to time and which, under the terms of the licence, bore the nomenclature, 'licence fee and royalty'.[5]

In December 1995, another 34 CMTS licences were granted based on the auction for 18 telecommunication circles for a period of 10 years. The 1995 licences mentioned that a cumulative maximum of up to 4.4 MHz in the 900 MHz bands shall be permitted to the licensees based on appropriate justification. Here again, there was no separate upfront charge for allocation of spectrum to the licensees who were also required to pay annual spectrum usage charges, which under the terms of the licence bore the nomenclature 'licence fee and royalty' subject to revision from time to time.[6]

In 1995, bids were also invited for Basic Telephone Service (BTS) licences with the licence fee payable for a 15-year period. Under the terms of the BTS licences, a licensee could provide fixed line as well as wireless basic telephone services. Six licences for providing BTS were granted in 1997–98 by way of auction through tender. The licence terms, inter alia, provided that based on the availability of the equipment for Wireless in Local Loop (WLL) in the world market, the

---
[5]https://indiankanoon.org/doc/37692759/
[6]Ibid.

spectrum in bands specified therein would be considered for allocation subject to the conditions mentioned therein. The licensees did not have to pay a separate upfront charge and those offering the basic wireless telephone service were required to pay annual spectrum usage charges, which were categorized as licence fee and royalty.[7]

In 1997, a significant step was taken by setting up an independent regulatory body, Telecom Regulatory Authority of India (TRAI) under the Telecom Regulatory Authority of India Act. It was established so that the telecom sector would be regulated in a balanced, fair and competitive manner. It was given statutory status by virtue of the Act.

The recommendations of the TRAI are not binding upon the Central Government. However, it is mandatory for the Central Government to seek TRAI's recommendations in respect of matters related to the issuance of a new licence to the service provider. The TRAI is required to forward its recommendation within a period of 60 days from the date of receipt of such recommendations.[8] Central Government can issue a licence to a service provider if no recommendations are issued from TRAI within the said period or within such period as may be mutually agreed upon between the Central Government and TRAI.[9]

If the Central Government, after having considered TRAI's recommendations, comes to a conclusion that such recommendations cannot be accepted or need modifications, it refers the recommendations back to TRAI for its

---

[7]Ibid.
[8]See Second Proviso to Section 11 (1)(a) of TRAI Act 1997.
[9]See Fourth Proviso to Section11 (1)(a) of TRAI Act 1997.

reconsideration. TRAI may, within 15 days from the date of receiving such a reference, forward its recommendations to the Government, after considering such reference. The Government is then required to take a final decision.[10]

In the case of Cellular Operators Association of India and others vs. Union of India and others,[11] the Supreme Court observed that:

> Due weightage has to be attached both to the Recommendations of TRAI which consists of an expert body as well as recommendations of GOT-IT.

Originally, TRAI was also empowered to adjudicate upon disputes among service providers or between the service providers and a group of consumers. Since there was a great demand for a separate dispute settlement mechanism, the TRAI Act, 1997 was amended in January 2000 and the Telecom Disputes Settlement and Appellate Tribunal (TDSAT) was established with both original and appellate jurisdictions.

Further, in consonance with the vision of the government, the tenth plan[12] indicated that the 'spectrum policy needs to be promotional in nature and revenue considerations play a secondary role. Pricing and allocation should ensure that available spectrum is utilized optimally. Spectrum pricing needs to be based on relative demand and supply over space and time in a dynamic manner. It also needs to ensure the introduction and promotion of spectrum efficient

---

[10]See Fifth Proviso to Section11 (1)(a) of TRAI Act 1997.
[11](2003) 3 SCC 186.
[12]2002–07

technology."[13] Also, as per the Cabinet decision 31 October 2003, it was stated that the spectrum pricing had to be finalized by DoT and the Ministry of Finance.[14]

Dawn of a New Era

By 1998, six service providers were in trouble and defaulted on payments. TRAI studied the situation and came to the conclusion that the average revenue per user (ARPU) was hugely over estimated and subscriber numbers much lower than projected. The New Telecom Policy of 1999 (NTP 1999) was brought into effect on the recommendations of a Group on Telecom (GoT) which had been constituted by the government to address inter alia the question of financial failure. It provided for fixed fee basis for licence and spectrum allocation. Essentially, the licence came bundled with 2.6 mW spectrum as start up. Any further spectrum could be sought on subscriber base with an escalating tariff for each additional spectrum. The industry found that to be unviable and lobbied vigorously for an alternative form of tariff. After considerable thought, the government of Prime Minister Atal Bihari Vajpayee switched to a revenue sharing model.

NTP 1999 provided that cellular mobile service providers (CMSP) would be granted a licence for 20 years on the payment of a one-time entry fee and licence fee in the form of revenue share. It also provided that BTS (fixed

---

[13]https://indiankanoon.org/doc/70191862/
[14]https://indiankanoon.org/doc/70191862/

service provider or FSP) licences for providing both fixed and wireless (WLL) services would also be issued for a period of 20 years on payment of a one-time entry fee and licence fee in the form of revenue share and prescribed charges for spectrum usage. The TRAI was to recommend the appropriate level of usage. The licensees, both cellular and basic, were also required to pay annual spectrum usage charges.[15]

Interestingly, on 22 July that year, NTP 1999 offered a migration package for migration from fixed licence fee to one-time entry fee and licence fee based on a revenue share regime to all existing licensees. This came into effect on 1 August. Under the migration package, the licence period for all the CMTS and FSP licensees was extended to 20 years from the date of issuance of licences.

In 1997 and 2000, the CMTS licences (PSU licences) were also granted in two and 21 circles to Mahanagar Telephone Nigam Ltd. (MTNL) and Bharat Sanchar Nigam Ltd. (BSNL), respectively. However, no entry fee was charged for these licences. The CMTS licences issued to both the PSUs granted GSM spectrum of 4.4 + 4.4 MHz in the 900 MHz band. The PSU licensees were also required to pay annual spectrum usage charges.[16]

In January 2001, based on TRAI's recommendation, the DoT issued guidelines for issuing CMTS licences for the fourth cellular operator. Based on tendering process structured as a 'Multistage Informed Ascending Bidding

---

[15]https://indiankanoon.org/doc/37692759/
[16]Ibid.

Process', 17 new CMTS licences were issued for a period of 20 years in the four metro cities and 13 telecom circles (the 2001 cellular licences).[17]

In the 2001 auction, spectrum, in a band other than 900 MHz, was auctioned for the first time in India. The government auctioned spectrum in the 1,800 MHz band using a three-stage auction process.

The 2001 licences required that the licensees pay a one-time, non-refundable entry fee as determined through auction, annual licence fee and annual spectrum usage charges. There was no separate upfront charge for spectrum allocation. In accordance with the terms of the tender document, the licence terms, inter alia, provided that a cumulative maximum of up to 4.4 + 4.4 MHz would be permitted. Further, based on usage, justification and availability, additional spectrum up to 1.8 + 1.8 MHz making a total of 6.2 + 6.2 MHz, may be considered for assignment, on a case-by-case basis, on the payment of additional licence fee. The bandwidth up to the maximum as indicated (4.4 MHz and 6.2 MHz, as the case may be) would be allocated based on the technology requirements (e.g. CDMA at 1.25 MHz, GSM at 200 KHz, etc.). Though frequencies assigned were not contiguous and not the same in all cases, efforts would be made to make available larger chunks to the extent feasible.

In 2001, the BTS licences were also issued for providing both fixed line and wireless basic telephone services on a continual basis (the 2001 Basic Telephone licences). The

---

[17]https://indiankanoon.org/doc/37692759/

licence terms, inter alia, provided that for wireless access system in local areas, not more than 5 + 5 MHz in 824–844 MHz paired with 869–889 MHz band shall be allocated to any basic service operator, including existing ones on FCFS basis. A detailed procedure for allocation of spectrum on FCFS basis was given in Annexure IX of the 2001 BTS licence. There was no separate upfront charge for spectrum allocation. Licensees were required to pay revenue share of 2 per cent of AGR earned from WLL subscribers as spectrum charges in addition to the one-time entry fee based on service-area and annual licence fee.

Following the 2001 auction, the government consciously abandoned the practice of auctions in favour of an administrative allocation model, under which the government would select companies that it deemed were best equipped to develop India's telecom infrastructure. The final allocation of 900 MHz took place in 2004 through this new model. This policy resulted in spectrum being allocated at far lower prices than had been done through auctions. For example, Reliance had paid ₹12.25 crore and ₹58.49 crore in 1995 for 4.4 MHz spectrum in West Bengal and Orissa respectively. In 2004, Airtel was allocated the same amount of spectrum in the same circles for ₹1 crore and ₹4.4 crore respectively.

In the following year, a subscriber-based criterion for CMTS was prescribed for allocation of additional spectrum of 1.8 + 1.8 MHz beyond 6.2 + 6.2 MHz with a levy of additional spectrum usage charge of 1 per cent of AGR. This allocation criterion was revised from time to time.

UAS Licence Regime

NTP 1999 was followed by the Unified Access Services (UAS) guidelines. These were the movements based on the experience so far. A telecom company that wishes to offer services in any of the 22 telecom circles in India must purchase a Unified Access Services Licence (UASL) to operate that circle. Licences are awarded by auctions and are valid for a period of 20 years, which can be extended by an additional 10 years once per licence per circle. Initially, a mobile network operator who was awarded a licence to operate in any of the telecom circles was allocated frequencies in that circle for a fixed time period. After the expiry of the licence, the company would have to bid to renew the licence.

On 27 October 2003, TRAI recommended a UASL regime. On 11 November, guidelines specifying the procedure for migration of existing operators to the new regime were issued. As per these guidelines, all applications for new access services licence would be in the UAS licence category. Later, based on TRAI's clarification of 14 November, the entry fee for new Unified licensees was fixed the same as the entry fee of the fourth cellular operator. Based on TRAI's further recommendations on 19 November, spectrum to the new licensees was to be given as per the existing terms and conditions relating to spectrum in the respective licence agreements.

After the FDI limit in the telecom sector was raised from 49 per cent to 74 per cent, revised guidelines for grant of UASLs were issued on 14 December 2005. These guidelines, inter alia, stipulated that licences shall be issued without any restriction on the number of entrants for provision of UAS in a service area. The applicant would be required to pay a one-time, non-refundable entry, annual licence fee as a percentage of AGR and spectrum charges on revenue share basis.

No separate upfront charge for allocation of spectrum was prescribed. Initial spectrum was allotted as per UASL conditions to the service providers in different frequency bands, subject to availability. Initially allocation of a cumulative maximum up to 4.4 + 4.4 MHz for TDMA-based systems or 2.5 + 2.5 MHz for CDMA-based systems was to be made. Spectrum not more than 5 + 5 MHz with respect to Code Division Multiple Access (CDMA) system or 6.2 + 6.2 MHz with respect to TDMA-based system was to be allocated to any new UAS licensee.

A new telecom policy announced by the government in 2011 delinked spectrum from licences. As a result, when an operator renews its licence it must also pay separately for spectrum.

It is important to remember that after the introduction of UAS in 2003 and until March 2007, 51 new UASLs were issued based on the policy of FCFS, on payment of the same entry fee that was paid for the 2001 cellular licences (the 2003–07 licences). The spectrum was also allocated on FCFS under a separate wireless operating licence on a case-by-case basis and subject to availability. The licensees had to pay annual spectrum usage charges as a percentage of AGR, since there was no upfront charge for spectrum allocation. The people who made the most noise about 2008 spectrum allocation conveniently forgot to explain why FCFS was introduced in the first place and why it was not changed over five years. In any case it is easy in hindsight to speak of notional losses unless one understands the conditions that prevailed in India in those years. In the rest of the world, particularly the developed economies, different forms of auctions were conducted as in UK, USA and Europe. But over the years, between 2001 and 2008, it was a challenge to maintain a level playing field for entrants at different stages.

On 28 August 2007, TRAI revisited the issue of new licences, allocation of spectrum, spectrum charges and entry fees, and issued its recommendations. In 2007 and 2008, the government issued Dual Technology licences where the terms of the existing licences were amended to allow licensees to hold a licence as well as spectrum for providing services through both GSM and CDMA network. The first amendment was issued in December 2007. All licensees who opted for Dual Technology licences paid the same entry fee which was an amount equal to the amount prescribed as entry fee for getting a new UASL in the same service area.

The amendment to the licence, inter alia, mentioned that initially a cumulative maximum of up to 4.4 + 4.4 MHz was to be allocated in the case of TDMA-based systems (at 200 kHz per carrier or 30 kHz per carrier). A maximum of 2.5 + 2.5 MHz was to be allocated in the case of CDMA-based systems (at 1.25 MHz per carrier), on a case-by-case basis, subject to availability. It was also mentioned that additional spectrum beyond the above stipulation could be considered for allocation after ensuring optimal and efficient utilization of the already allocated spectrum taking into account all types of traffic and guidelines/criteria prescribed from time to time. However, spectrum not more than 5 + 5 MHz in respect to CDMS system and 6.2 + 6.2 MHz in respect to TDMA-based system was to be allocated to the licensee. There was no separate upfront charge for spectrum allocation. However, the Dual Technology licensees were required to pay spectrum usage charges in addition to the licence fee on revenue share basis as a percentage of AGR. Spectrum to these licensees was allocated from 10 January 2008 onwards.

In case of spectrum allotted beyond 6.2 MHz, the frequency allocation letters issued by DoT from May 2008 onwards, mentioned that allotment of spectrum was subject to pricing as determined in future by the government and the outcome of the court orders. However, annual spectrum usage charges were levied on the basis of AGR, as per the quantum of spectrum assigned.

In 2008, 122 new 2G UASLs were granted to telecom companies on an FCFS basis at the 2001 price. Letters of Intent (LoI) were issued for 122 licences for providing 2G services on or after 10 January 2008, against which licences (the 2008 licences) were subsequently issued. However, it is these licences that were quashed, by the Supreme Court judgement of 2 February 2012 in Centre for Public Interest Litigation vs. Union of India ([2012] 3 SCC 1).

3G Spectrum Auction

In 2010, 3G and 4G telecom spectrum were auctioned in a highly competitive bidding. The terms of the auction stipulated that for successful new entrants, a fresh licence agreement would be entered into and for existing licensees who were successful in the auction, the licence agreement would be amended for use of spectrum in the 3G band. According to the terms of the amendment letter, 3G spectrum would stand withdrawn if the licence was terminated for any reason.

The private companies that participated in the auction were Airtel, Aircel, Idea, Reliance Communications, S Tel, Tata Teleservices and Vodafone Essar. State-owned telecom companies BSNL and MTNL were also awarded spectrum, although they did not have to participate in the auction. BSNL paid the government ₹101.87 billion (equivalent to US$2.4 billion) for spectrum in 20 circles and MTNL got spectrum for 3G services in 2 circles—Delhi and Mumbai.

The winners were awarded spectrum in September that year, and Tata Docomo was the first private operator to launch 3G services in India. The government earned ₹677 billion (equivalent to US$16 billion) from the 3G spectrum auction and the broadband wireless spectrum auction generated a revenue of ₹385 billion (equivalent to US$9.0 billion) for a total revenue of ₹1,062 billion (equivalent to US$25 billion) from both auctions.

Based on the recommendations of TRAI dated 11 May 2010, followed by further clarifications and recommendations, the government in February 2012, prescribed the limit for spectrum assignment in the metro service areas as 2 MHz × 10 MHz/2 MHz × 6.25 MHz and in rest of the service areas as 2 MHz × 8 MHz/2 MHz × 5 MHz for GSM (900 MHz, 1,800 MHz band) and CDMA (800

MHZ band) respectively. This was subject to the condition that the licensee could acquire additional spectrum beyond the prescribed limit in the open market should there be an auction of spectrum subject to the further condition that total spectrum held by it does not exceed the limits prescribed for merger of licences, i.e. 25 per cent of the total spectrum assigned in that service area by way of auction or otherwise. The limit for CDMS spectrum was 10 MHz.

In view of the fact that spectrum may need to be allocated to individual entities from time to time in accordance with the criteria laid down by the government, such as subscriber base, availability of spectrum in a particular circle, inter se priority depending on whether the spectrum comprises the initial allocation or additional allocation, etc., it may not always be possible to conduct an auction for spectrum allocation.

In view of the above, the auctioning of spectrum in the 2G bands could result in a situation where none of the licensees using the 2G bands of 800 MHz, 900 MHz and 1,800 MHz would have paid any separate upfront fee for the allocation of spectrum.

Far-reaching Implications

The 2G spectrum allocation case inevitably brought policy changes due to a once-bitten-twice-shy approach, such as a shift to allotting airwaves through auctions, and a consolidation in the sector—which is still a work in progress. There were 16 operators offering mobile services in the country before the Supreme Court cancelled 122 telecom licences in February 2012. The present number is 11 and could soon become five, if the proposed mergers and acquisitions go through. This consolidation spree is considered to be one of the major fallouts of the decade-old allegations over irregularities in the issue of telecom licences and 2G spectrum.

The 2G trial court in 2017 may have acquitted all the accused in the criminal cases pertaining to the alleged irregularities, but the policy changes and decisions taken in the aftermath of the allegations have impacted India's telecom sector in a number of ways. These include the Centre's shift of stance to allocate spectrum by letting the market decide the price of the natural resource instead of allotting it to telecom operators bundled with their licences. The Centre's pursuit of higher revenues through auctions, which is seen to be a direct result of the CAG's observations on the loss of revenue through administrative allocation, has also resulted in spectrum becoming costlier, and the consequent debt situation—the roots of which pre-date the current non-performing asset (NPA) scenario shadowing the country's financial services sector today. According to estimates, the telecom sector is in a combined debt of over ₹5 lakh crore. In the 2016 auction, despite a clear lack of intent from the industry, the highly priced spectrum in the 700 MHz frequency was put under the hammer. In its internal pre-auction estimates, even the DoT did not expect the spectrum to be fully sold. Throughout the auction, not a single bid by any operator was placed.

Furthermore, on account of the high debt in the sector, earlier this year, the Reserve Bank of India (RBI) red-flagged the telecom industry and asked banks to review their exposure to the sector. An inter-ministerial group was formed, headed by DoT Secretary Aruna Sundararajan, to suggest measures for reducing financial stress in the sector.

Even as the Supreme Court's 2012 order pertained to a different aspect of the case, the decision had an impact on banks that considered writing off their loans given to licensees since lenders were exposed to the sector in the form of loans provided as guarantees for acquiring the licences.

The FCFS policy adopted by the UPA government to allocate licences was in fact a part of the NTP 1999, which gave priority to increase in tele density in the country. This was reflected in the telecom subscription data in the report issued by the CAG. The total number of wireless connections in the country grew from 2.28 crore to 26.19 crore in 2008, and to 58.43 crore in 2010. As of 31 October 2017, the total number of wireless subscribers was 117.82 crore.

As per the CAG report, the government, by allocating the 122 licences for 2G services and 35 dual technology licences in 2008, realized revenues of ₹12,386 crore. Compared with this, in the six editions of spectrum auctions from 2010 to 2016, operators have committed to spectrum worth almost ₹2.63 lakh crore, of which spectrum worth nearly ₹1.76 lakh crore has been auctioned by the current NDA government.

The 2G auction of spectrum of November 2012 turned out to be disappointing—just two days, as compared to the 34 days that the 3G spectrum auction took, and it ended raising just ₹9,407 crore against a target of ₹40,000 crore with only five participants, of which only two were among those whose licences had been cancelled. What is more, only 42.37 per cent of the airwaves that were put up for sale were successfully auctioned. This was achieved despite artificial shortage having been created by offering only a part of the total spectrum returned under the court order. Consequently, further rounds of auction had to be conducted and DoT lowered reserve fees after each failed round. When it started auctions for 1,800 MHz spectrum in November 2012, it kept the reserve fee at an eye-popping ₹14,000 crore for a 5-MHz pan-India block. There were no bids in the four major circles of Delhi, Mumbai, Karnataka and Rajasthan. Worse was the fact that there were no bids received in the 800 MHz spectrum.

For March 2013 auctions, the DoT lowered the reserve fee for the four failed circles by as much as 30 per cent. But still, bidders found the reserve fee to be too high, and there was no bidding. It was a complete disaster. In the 800 MHz band, despite a 50 per cent lowering in reserve fees, there was only one bidder—Sistema Shyam—which picked up eight circles for ₹3,639 crore. For the highly efficient 900 MHz spectrum—being 'refarmed' for the first time—there were no bids at all.

In the 2014 round, the DoT was forced to reduce the reserve fees further. The 1,800 MHz reserve fee was lowered by a further 26 per cent and the 900 MHz one by 50 per cent. It is a result of these reductions that the government said that it would earn ₹61,162 crore from the 2G spectrum auction that ended after 68 rounds of bidding over 10 days. Major telecom companies Airtel and Vodafone bagged spectrum in the crucial 900 MHz band in important markets like Delhi, Mumbai and Kolkata.

Auctions in November 2012 and March 2013 flopped as most bidders stayed away, complaining that the floor bid prices were too high. Eight bidders applied to participate in the 2014 auction after the government sharply cut auction reserve prices. At ₹61,162 crore, the government's total revenue from the auction was much higher than its initial estimate of about ₹41,000 crore. These licences will be valid for a period of 20 years. The companies need to pay only a quarter to a third of the winning auction price upfront and the remainder through to 2026.

It is possible that the trauma caused by the cancellation saga has finally come to an end and the mobile telephony sector is well on the path of recovery and normalcy. But have we learnt any lessons except that public life in India is inherently perilous? Political ambitions are ruthless in the consequences for the opponent; reputations are destroyed without remorse and 'sorry' is a word reserved for moments of compulsion rather than realization. Demonetization, for instance, inflicted on hapless citizens has the same element of callous disregard for the physical and psychological price that people pay for politicians' ambitions.

3
2G CASE: THE BARE FACTS

To understand the core of the 2G controversy, one has to reflect briefly on the structure of the corporate engagement. At the time of issuing of 2G licences, there were two groups of service providers. The first, comprising service providers of GSM, included Airtel, Aircel, Vodafone, Idea, while the second was AUSPI (Association of Unified Telecom Service Providers of India) using CDMA—essentially Sistema and Reliance Telecom. It must be recalled that there was intense tension between the two groups with vigorous lobbying. Essentially, the GSM group sought further spectrum to what they already had acquired under the previous regime without extra payment and that too on the terms of subscriber base more favourable to them. Furthermore, they had a problem with permission for dual technology that the CDMA service providers were to be given on the assumption of technology neutrality accepted by TRAI in its recommendations.

The TRAI recommendations were made in August 2007. The 200-page report, in addition to playing a catalytic role in development, sought to balance accepting of 'legacy and level playing field' with 'social good or consumer interests. The Regulator pitched for auction in 3G and beyond but keeping the existing situation in mind opted to continue FCFS for UAS 2G licences:

> The allocation of spectrum is after the payment of entry fee and grant of license. The entry fee as it exists today is, in fact, a result of the price discovered through a market based mechanism applicable for the grant of license to the 4th cellular operator...It is in this background that the Authority is not recommending the standard options pricing of spectrum, however, it has elsewhere in the recommendation made a strong case for adopting auction procedure in the allocation of all other spectrum bands except 800, 900 and 1800 MHz (para 2.73).

It is interesting to note that the Chairman of TRAI was none other than Nripendra Mishra, presently principal secretary to PM Modi.

It is surprising therefore, that the CAG, CVC and CBI, not to mention the Supreme Court should have so easily persuaded themselves that not putting 2G spectrum to auction was a grievous infraction of the law and by implication, an act of corruption causing enormous loss to the country's exchequer.

Much was made of a letter dated 2 November 2007, whereby Prime Minister Dr Manmohan Singh reportedly advised Telecom Minister A. Raja to allot 2G spectrum transparently and revise the licence fee.[18] Raja claims to have explained each point to the PM, although in the exchange of letters (one lot having crossed in transmission) it is difficult to say that many of Dr Singh's recommendations were bypassed.

In another letter that month, the Ministry of Finance expressed procedural concerns to the DoT regarding how the rate of ₹1,650 crore determined in 2001 was applied to licences given in 2007. The answer was inevitably that NTP 1999, UASL Guidelines, TRAI recommendations and the Cabinet decision of 2003 were conclusive in the Ministry's decision of not making any change in the entry fee. In both cases, selective reading of the contents gives an impression which is far from the truth. But clearly the detractors of the government were not interested in the truth; their intent was to state the allegations in a manner that culpability would become conclusive without a fair trial. Raja's complaint was that despite his having done everything by the book and keeping the PMO informed, there was little effort to aggressively defend him against the hostile mob. There can be but only one explanation: under severe and vicious attack, our instinct was to protect the top leadership and keep it above the daily slugfest in order to keep the government intact. We knew that the real object of the Opposition was to hurt where it really mattered and the individual ministers were only the stepping stones to the ultimate strike. Whatever

---

[18]Andimuthu Raja, *2G Saga Unfolds*, Har-Anand Publication, Annexure-II, pp. 200–1.

damage control could be undertaken besides protecting the office of the prime minister was done to the best of our ability keeping in mind the charged atmosphere. The Opposition had known for long that the PM's unimpeachable integrity had been a major hurdle for their plans and when nothing else worked they resorted to profane propaganda against him as well.

Besides the entry fees, the CAG and the prosecution believed there was serious malfeasance in the cut-off date for consideration of applications which was moved forward from 1 October 2007 (chosen over 10 October 2007) and then to 25 September (the day after the Press Release about the cut off was made). On 25 September, the DoT announced on its website that applicants filing between 3:30 p.m. and 4:30 p.m. that day would be granted licences and the spectrum would then be released on priority of compliance. Thus what was 'first come first serve' became 'first comply first serve'. Raja's complaint has been that even though his colleagues resisted allegations regarding policy, they self-consciously conceded ground on incorrect application of policy. The 2G case thus became a test case of our inability to understand the adversaries' strategy and becoming defensive to cut losses. Sadly in this, much respected Ghulam Vahanvati, Attorney General of India, became a target and suffered immeasurable anguish for having to answer summons of the trial court.

Finally it was alleged that although the corporations that were ineligible—Swan Telecom and Unitech (Uninor)—were granted licences for ₹1,658 crore and ₹1,537 crore respectively. The former sold 67.25 per cent equity for ₹6,200 crore to Norway-based Telenor and the latter sold 45 per cent to UAE-based Etisalat for ₹4,200 crore. This was a critical ingredient of calculation of presumptive loss by the CAG in addition to the figure of 3G auction prices. It is noteworthy and will be examined in some detail later that inter alia these allegations did not find favour with the trial court.

An interesting aspect of the case was that no money trail was found, even when seriously investigated and yet, the Prevention of Corruption Act was resorted to. The only case of money transfer was ₹200 crore passing from Shahid Balwa of DB Realty to Kalaignar TV controlled by the DMK.

The details of companies which received 2G licences during Raja's term as telecom minister (the licences were all later cancelled by the Supreme Court) are as follows:

Company	Telecom Regions	Number of Licences
Adonis Projects	Haryana, Himachal Pradesh, Jammu & Kashmir, Punjab, Rajasthan, Uttar Pradesh (East)	6
Nahan Properties	Assam, Bihar, North-East, Orissa, Uttar Pradesh (East), West Bengal	6
Aska Projects	Andhra Pradesh, Kerala, Karnataka	3
Volga Properties	Gujarat, Madhya Pradesh, Maharashtra	3
Azure Properties	Kolkata	1
Hudson Properties	Delhi	1
Unitech Builders & Estates	Tamil Nadu (including Chennai)	1

Unitech Infrastructures	Mumbai	1
Loop Telecom	Bihar, Gujarat, Himachal Pradesh, Kerala, Kolkata, Punjab, Rajasthan, Uttar Pradesh, West Bengal, Andhra Pradesh, Delhi, Haryana, Karnataka, Maharashtra, Odisha, Tamil Nadu (including Chennai), Assam, Jammu & Kashmir, Madhya Pradesh	21
Datacom Solutions	Andhra Pradesh, Assam, Bihar, Gujarat, Haryana, Himachal Pradesh, Jammu & Kashmir, Karnataka, Kerala, Kolkata, Madhya Pradesh, Maharashtra, Odisha, Rajasthan, Tamil Nadu (including Chennai), Uttar Pradesh, West Bengal, Delhi, Mumbai	21
Shyam Telelink	Madhya Pradesh, Kerala, Kolkata, Punjab, Uttar Pradesh, West Bengal, Andhra Pradesh, Delhi, Haryana, Karnataka, Maharashtra, Odisha, Tamil Nadu (including Chennai), Assam, Jammu & Kashmir, North-East	17
ShyaniTelelink	Mumbai, Bihar, Gujarat, Himachal Pradesh	4

Swan Telecom	Andhra Pradesh, Gujarat, Haryana, Karnataka, Kerala, Maharashtra, Punjab, Rajasthan, Tamil Nadu (including Chennai), Uttar Pradesh, Delhi, Mumbai	13
Allianz Infratech	Bihar, Madhya Pradesh	2
Idea Cellular	Assam, Punjab, Karnataka, Jammu and Kashmir, North-East, Kolkata, West Bengal, Odisha, Tamil Nadu (including Chennai)	9
Spice Communications	Delhi, Andhra Pradesh, Haryana, Maharashtra	4
S Tel	Assam, Jammu and Kashmir, Odisha, North-East, Bihar, Himachal Pradesh	6
Tata Teleservices	Jammu and Kashmir, Assam, North-East	3

Although several of these successful bidders were found to be at fault by the Supreme Court, at no stage did any unsuccessful party seriously claim to have been denied just deserts or their entitlement due to unfair or dishonest allocation of spectrum.

The Key Players

The questions about the licences drew attention to three groups: politicians and bureaucrats in authority; corporations; and professionals who allegedly mediated between the politicians and corporations. Accordingly, the following names blink on the dashboard of controversy:

A. Raja

Representative of UPA major alliance partner, the four-time DMK Member of Parliament who had won the 15th Lok Sabha elections from Nilgiris constituency in Tamil Nadu, Raja was Union Minister of State for Rural Development in 1999, Health and Family Welfare in 2003, and Union Cabinet Minister for Environment and Forests in 2004. He held the Communication and Information Technology portfolio in 2007 and 2009. The CBI charge sheet alleged that Raja conspired with the other accused and arbitrarily refined the FCFS policy to ensure that Swan and Unitech received licences. Furthermore, instead of auctioning 2G spectrum, he sold it at the 2001 rate.

As a result, he was charged with criminal breach of trust by a public servant (Section 409), criminal conspiracy (Section 120B), cheating (Section 420) and forgery (Sections 468 and 471) and was booked under the Prevention of Corruption Act (PCA) for accepting illegal gratification although no serious effort was made to establish that during investigation and trial.

M.K. Kanimozhi

Daughter of the five-time Chief Minister of Tamil Nadu M. Karunanidhi, M.K. Kanimozhi is DMK MP, representing Tamil Nadu in the Rajya Sabha. According to the CBI charge sheet, Kanimozhi owns 20 per cent of the family-owned Kalaignar TV; her stepmother, Dayalu Ammal, owns

60 per cent of the channel. The CBI alleged that Kanimozhi was the 'active brain' behind the channel and conspired with Raja to coerce DB Realty co-founder Shahid Balwa to funnel ₹200 crore to Kalaignar TV although as a corporate loan. Raja advanced the channel's cause, facilitating its registration with the Ministry of Information and Broadcasting and adding it to DTH operator TATA Sky's line-up. Kanimozhi was also charged with tax evasion by the Income Tax Department in Chennai. She was charged with criminal conspiracy to cause criminal breach of trust by a public servant and criminal conspiracy (Section 120B), cheating (Section 420) and forgery (Sections 468 and 471), and booked under the Prevention of Corruption Act.

Given her position in the DMK family and being an articulate and energetic public figure, Kanimozhi was obviously influential and at the very centre of decision-making in her party. But there was no explicit evidence of her masterminding any unlawful gratification.

Bureaucrats

A number of bureaucrats were named in the CBI charge sheet filed in the Special Court.

Siddharth Behura

Competent and pleasant IAS officer from the UP cadre belonging to a family of civil servants, Siddharth Behura became Telecom Secretary just before the licences were granted. According to the CBI charge sheet, Behura conspired with Raja and several others mentioned in this list.

It was alleged that when the application deadline time was declared to be 3:30 p.m.–4:30 p.m., Behura closed counters early to block telecom companies other than the favoured ones. However, it was not made clear as to how this affected anyone adversely.

He was charged with criminal breach of trust by a public servant (Section 409), criminal conspiracy (Section 120B), cheating (Section 420) and forgery (Sections 468 and 471) and was booked under the Prevention of Corruption Act for allegedly accepting illegal gratification although once again, no money trail was found.

R.K. Chandolia

Raja's private secretary when the licences were granted, Chandolia, as per the CBI charge sheet, conspired with Raja and several others. When the deadline for the submission of the applications was altered from 3:30 p.m. to 4:30 p.m., Chandolia joined Behura in shutting down counters to physically block some telecom companies. He was charged with criminal conspiracy (Section 120B), cheating (Section 420) and forgery (Sections 468 and 471). He too was booked under the Prevention of Corruption Act.

Business Executives

The executives accused in the CBI charge sheet included:

Sanjay Chandra

The former managing director of Unitech Wireless, Chandra was charged with criminal conspiracy (Section 120B), cheating (Section 420) and forgery (Sections 468 and 471) being booked under the Prevention of Corruption Act. During the trial, the then CBI prosecutor A.K. Singh was implicated in a taped conversation sharing legal strategy and privileged information with Chandra. After spending several months in prison, he was granted bail by the Supreme Court through the celebrated liberty-oriented judgement authored by Justice H.L. Duttu who was later to be Chief Justice of India.

Gautam Doshi

Although there was much speculation that Anil Ambani might find himself in the prosecutor's net, there was a great sigh of relief that the charges were restricted to his senior employees. The managing director of Reliance Anil Dhirubhai Ambani Group, Doshi was charged with criminal conspiracy (Section 120B), cheating (Section 420) and forgery (Sections 468 and 471). He was also booked under the Prevention of Corruption Act.

Hari Nair

Like the other business executives, Nair, senior vice-president of Reliance Anil Dhirubhai Ambani Group, was also charged with criminal conspiracy (Section 120B), cheating (Section 420) and forgery (Sections 468 and 471). He was booked under the Prevention of Corruption Act too.

Surendra Pipara

Like his colleague, this senior vice-president of the Reliance Anil Dhirubhai Ambani Group was charged with criminal conspiracy (Section 120B), cheating (Section 420) and forgery (Sections 468 and 471) and was also booked under the Prevention of Corruption Act.

Vinod Goenka

Goenka was the managing director, DB Realty and Swan Telecom. He was charged with criminal conspiracy (Section 120B), cheating (Section 420), forgery (Sections 468 and 471) and fabrication of evidence (Section 193), and was also booked under the Prevention of Corruption Act.

Shahid Balwa

The corporate promoter of DB Realty and Swan Telecom, Balwa was also charged with criminal conspiracy (Section 120B), cheating (Section 420), forgery (Sections 468 and 471) and fabrication of evidence (Section 193), being additionally booked under the Prevention of Corruption Act.

Asif Balwa and Rajiv Agarwal

Director of Kusegaon Fruits and Vegetables, Balwa was charged with criminal conspiracy (Section 120B), cheating (Section 420), forgery (Sections 468 and 471) and fabrication of evidence (Section 193); he too was booked under the Prevention of Corruption Act.

Also a Director of the same firm like Balwa, Agarwal was charged with criminal conspiracy (Section 120B), cheating (Section 420), forgery (Sections 468 and 471) and fabrication of evidence (Section 193) and booked under the Prevention of Corruption Act.

Sharath Kumar

The managing director of Kalaignar TV, he was charged with criminal conspiracy (Section 120B), cheating (Section 420), forgery (Sections 468 and 471) and fabrication of evidence (Section 193). Kumar was booked under the Prevention of Corruption Act.

Ravi Ruia, Anshuman Ruia and Vikas Saraf

The vice-chairman of the Essar Group, Ravi Ruia was charged with criminal conspiracy (Section 120B) and cheating (Section 420).

Anshuman Ruia, a director with the Essar Group, was also charged with criminal conspiracy (Section 120B) and cheating (Section 420). Saraf was director of strategy and planning of the group and was charged with criminal conspiracy (Section 120B) and cheating (Section 420 of the Indian Penal Code).

I.P. Khaitan and Kiran Khaitan

The corporate promoter of Loop Telecom I.P. Khaitan was charged with criminal conspiracy under (Section 120B) and cheating (Section 420). Similarly, Kiran Khaitan, another corporate promoter, was charged with criminal conspiracy (Section 120B) and cheating (Section 420). The Khaitans were accused of violating the 'substantial equity' restriction imposed on association with a company already in possession of a licence.

Karim Morani

Karim Morani is the corporate promoter and director of

Cineyug Films. According to the Income Tax Department charge sheet, Morani owned Cineyug Films and was a part of the route used by Shahid Balwa to funnel ₹200 crore ($31 million) illegally to Kalaignar TV. DB Realty corporate promoters Shahid Balwa and Vinod Goenka transferred the money to Kusegaon Fruits and Vegetables, where Balwa's younger brother Asif was a director. Kusegaon then transferred ₹200 crore to Cineyug Films, and Morani transferred it to Kalaignar TV. He was charged with criminal conspiracy to cause criminal breach of trust by a public servant, criminal conspiracy (Section 120B), cheating (Section 420), forgery (Sections 468 and 471) and fabrication of evidence (Section 193), and was booked under the Prevention of Corruption Act.

Corporations

Several companies were also named in the CBI charge sheet. Unitech Wireless was charged with criminal conspiracy (Section 120B), cheating (Section 420) and forgery (Sections 468 and 471). Reliance Telecom was charged with criminal conspiracy to cause criminal breach of trust by a public servant, criminal conspiracy (Section 120B) and cheating (Section 420). Swan Telecom was also charged with criminal conspiracy to cause criminal breach of trust by a public servant, criminal conspiracy (Section 120B) and cheating (Section 420). The other companies named in the charge sheet included Loop Telecom, Loop Mobile India and Essar Tele Holding and Essar Group (corporate parent of Essar Tele Holding).

Bolt from the Blue: The CAG Report

Vinod Rai, a Kerala cadre IAS officer, was chosen by the UPA to be CAG over several other likely candidates at a late night intervention by the finance minister. Nothing about his record would have suggested that he would end up virtually destroying the government. He was friendly and communicative in the initial period of his tenure but with the Report he submitted, he became a hot potato.

Most people went by the media reporting of the contents of the CAG Report and the stray statements of political rivals of the government. Few had access to the Report or would have seriously attempted to read it. The critical excerpts of the Report make interesting reading:

> The telecom sector had witnessed dynamic and rapid transition. It had been subject to audit and a report titled 'Package of Concessions Given to Cellular Mobile Operators' was presented to Parliament in May 2000.[19] A further review of the 'Revenue Management in the Department of Telecommunications' was also undertaken by this office in 2004–05.[20] This review mainly focused on the system of collection and

---

[19]https://www.thehindubusinessline.com/todays-paper/Executive-summary-of-CAG-report/article20024712.ece.; http://www.thehindu.com/migration_catalog/article15671228.ece/BINARY/Executive%20Summary

[20]https://www.thehindubusinessline.com/todays-paper/Executive-summary-of-CAG-report/article20024712.ece; http://www.thehindu.com/migration_catalog/article15671228.ece/BINARY/Executive%20Summary

accounting of licence fee and spectrum charges from the licensees. The Report based on this review was presented to the Parliament in May 2006.[21] In January 2008, DoT issued 120 new licences for UAS on the same day at prices that had been discovered in 2001. The issue of 120 licences in just one day and at a price discovered in 2001 drew the attention of media, Parliament and informed members of the civil society. Questions were raised regarding the transparency in the allocation process and the failure in maximizing revenue generation from the allocation of spectrum, which is a national asset. The department received innumerable references from MPs and other sources repeatedly, questioning the allocation process and the price. The claim in each such reference was that ineligible applicants seemed to have been granted licences and at a price which appeared far below than the perceived appropriate market price in 2008. It was in this context that the DoT felt that there was sufficient justification to review the entire process of issue of licences, award of spectrum and the implementation of the UAS regime. The need for doing so was further justified as six years had passed since the introduction of the UAS regime in 2003. While accepting the Government's prerogative to formulate the policy of UASL, it was felt that an in-depth examination of implementation of such policy needed to be done.

---

[21]Ibid.

Amongst the critical findings of the Report on different aspects of the spectrum allocation process, the following points stand out:

Gaps in Policy Implementation

In August 2003, TRAI had submitted a Report recommending a roadmap for allocation of licences. This Report formed the basis for the UAS policy approved by the Council of Ministers in October 2003. The implementation of UASL regime was to be carried out in two phases with first phase of six months assigned for migration of already existing Basic Service Operators (BSOs) and Cellular Mobile Service Operators (CMSOs) to the new regime. The entry fee for migration of BSOs was determined as the fee equal to what was paid by the fourth cellular operator introduced through multistage bidding process in 2001. CMSOs were not required to pay any entry fee for migrating as they had already entered the market through a bidding process and thus paid a market determined price. The second phase was to start after the first phase in which a Unified Licencing regime, with a nominal entry fee for the licence with the spectrum being charged separately, was envisaged.

However, audit examination reveals that the DoT did not implement the licencing regime as approved by the Cabinet and implemented only the first phase of the policy, overlooking the second phase. In the actual implementation, the interim stage of implementation seems to have become the final destination. This appears to have become the underlying factor, quite erroneously, to value the spectrum in 2008 at 2001 prices.

An important objective of this policy decision to delink the prices of spectrum from the issue of licence and devise an efficient allocation formula for spectrum along with an appropriate price remained unachieved. Ministry of Finance was authorized by the Cabinet decision of 2003 to participate in the discussion for efficient allocation of spectrum and price fixation but DoT decided not to associate the Ministry of Finance.

As a consequence of such lacunae in the implementation of the policy laid down by the Council of Ministers in 2003, the issuance of licences in 2008 alongwith allocation of spectrum has been done by DoT at prices determined in 2001 which were based on a totally nascent market, despite their sector witnessing substantial transformation and manifold growth. The issue was never placed before Cabinet for a review.[22]

Telecom Commission Was Not Consulted

From a scrutiny of the records and information made available, it appears that the High Powered Telecom Commission, which also includes part time members from the Ministry of Finance, IT and Planning Commission,was not apprised of the TRAI

---

[22]https://www.thehindubusinessline.com/todays-paper/Executive-summary-of-CAG-report/article20024712.ece; Paras 3.1, 3.2, 3.3, http://www.thehindu.com/migration_catalog/article15671228.ece/BINARY/Executive%20Summary

recommendations of August 2007 and hence, was not afforded an opportunity to deliberate on the merits of the TRAI recommendations. It is also seen that the High Powered Telecom Commission was not even consulted at the time of grant of 122 UASL in 2008.[23]

Views and Concerns of Ministry of Finance Overruled

It was noted in Audit that DoT managed to keep the issue of spectrum pricing outside the purview of the GoM. The GoM's role in December 2006 was confined to issues concerning spectrum vacation. The Term of References (ToRs) left out the other two issues of efficient allocation and pricing, while all three were pronounced in the policy decision of 2003. Thus, by getting the spectrum pricing issue deleted from the ToR, the DoT completely side-tracked the pricing issues.[24]

It has also been revealed in the course of audit that the Ministry of Finance in November 2007 had questioned the sanctity of continuing with the price determined way back in 2001 without any index at current valuation. The Ministry had sought a review of the matter. This advice of the Ministry of Finance was overlooked by the DoT ostensibly on the basis of a four-year old Cabinet decision (October 2003) on the premise that it was authorized to calculate the entry fee for licences as per the recommendations of TRAI

---

[23](Paras 4.2, 4.5), http://www.thehindu.com/migration_catalog/article15671228.ece/BINARY/Executive%20Summary
[24]https://www.thehindubusinessline.com/todays-paper/Executive-summary-of-CAG-report/article20024712.ece

in 2003. DoT maintained that 'spectrum pricing was within the normal work carried out by them.'[25]

Advice of Ministry of Law and Justice Was Ignored

In October 2007, at its own initiative, the DoT requested the Ministry of Law and Justice to obtain and communicate the opinion of the Attorney General/Solicitor General of India to enable the DoT to handle an unprecedented rush of applications in a fair and equitable manner which would be legally tenable. The Ministry of Law, at the level of the Minister, opined that in view of the importance of the case and the various options which seem to have emerged, it was necessary that the whole issue be first considered by an Empowered Group of Ministers (EGoM) and in that process legal opinion of the Attorney General can be obtained. Surprisingly, this opinion, which the DoT had sought on its own volition, was felt to be 'out of context' at the level of the Ministry of Communications and Information Technology (MoC&IT) and hence the benefit of a discussion in the EGoM was also forgone. Thus, such important decisions seem to have been taken in DoT without the issues being deliberated and discussed at an inter-ministerial forum.[26]

---

[25]https://www.thehindubusinessline.com/todays-paper/Executive-summary-of-CAG-report/article20024712.ece
[26]https://www.thehindubusinessline.com/todays-paper/Executive-summary-of-CAG-report/article20024712.ece; Para 4.3, http://www.thehindu.com/migration_catalog/article15671228.ece/BINARY/Executive%20Summary

Hon'ble Prime Minister's Suggestions Were Not Followed

In November 2007, the Hon'ble Prime Minister wrote to MoC&IT and expressed concern that in the backdrop of the inadequate spectrum and the unprecedented number of applications received for fresh licences, spectrum pricing through a fair and transparent method of auction for revision of entry fee, which is currently benchmarked on an old figure, needs to be reconsidered. This advice of the Prime Minister evoked an immediate response from the MoC&IT who, on the same day, replied that the issue of auction of spectrum was considered by the TRAI and the Telecom Commission and it was not recommended by them as the existing licence holders had already got spectrum upto 10 Mhz per circle without any spectrum charge. MoC&IT further informed that his ministry has come to the conclusion that it will be unfair, discriminatory, arbitrary and capricious to auction spectrum to new applicants as it will not give them a level-playing field. He had thus justified the allotment of spectrum to a few new operators in 2008 without reconsidering the old entry fee discovered in 2001, ignoring the advice of the Prime Minister.[27]

---

[27]https://www.thehindubusinessline.com/todays-paper/Executive-summary-of-CAG-report/article20024712.ece; Para 4.4, http://www.thehindu.com/migration_catalog/article15671228.ece/BINARY/Executive%20Summary

Arbitrary Changes by DoT in the Cut-Off Date

The TRAI report of August 2007 had recommended 'no cap' on the number of licences in any service area. Despite this recommendation of TRAI, the DoT issued a Press Release on 24 September 2007 stating that applications for issue of licences would be accepted only upto 1 October 2007. This action, in effect, conveyed fixation of an artificial cap in the number of licences to be awarded. However, in its response (July 2010) to the report issued to the Ministry (July 2010), the Ministry has stated that it accepted the recommendation of 'no cap' by the TRAI in October 2007. It seems that the Ministry, by issuing the press release in advance in September 2007 had, in effect, circumvented the recommendation of TRAI by taking an action counter to the recommendation and its acceptance by DoT in October 2007. To further compound the earlier decision, of restricting consideration of applications received upto 1 October 2007, the DoT further advanced this date to restrict issuance of Letters of Intent (LoIs) only to applications received by 25 September 2007. This was ostensibly to avoid legal implications in view of the shortage of spectrum for GSM services.[28]

---

[28]https://www.thehindubusinessline.com/todays-paper/Executive-summary-of-CAG-report/article20024712.ece; Paras 4.1.2, 4.6, http://www.thehindu.com/migration_catalog/article15671228.ece/BINARY/Executive%20Summary

FCFS Policy Was Not Followed

The First Come First Serve (FCFS) policy earlier internally adopted in DoT for allocation of spectrum was then extended for issue of new UASL. Under this policy, all applications are registered in the Central Registry Section of DoT where date of receipt and serial numbers are posted on it. Priority of applications is determined based on this date of receipt in the Central Registry. In a communication dated 2 November 2007, the MoC&IT had even confirmed to the Prime Minister that the processing of applications was to be on the FCFS basis. However, audit found that DoT deviated even from the FCFS policy in letter and spirit. The applications submitted between March 2006 and 25 September 2007 were issued the LoIs simultaneously on a single day, viz. 10 January 2008. A notice was issued through a press release giving less than an hour to collect the same. This decision to issue LoIs simultaneously to all applicants was taken at the level of the Minister. As per the FCFS policy being followed, those who were issued LoIs were given 15 days to fulfil the conditions. This included submission of a Performance Bank Guarantee (PBG) and a Financial Bank Guarantee (FBG). By changing the FCFS criteria, some licensees, who could proactively anticipate such procedural changes were ready with the Demand Drafts drawn on dates prior to the notification of cut-off date by DoT and could avail the benefit of first right to allocation of spectrum, having jumped the queue.

The entire process followed lack of transparency and objectivity and has eroded the credibility of DoT.[29]

Issue of Licence to Ineligible Applicants

The process followed by the DoT for verification of applications for UASL for confirming their eligibility lacked due diligence, fairness and transparency leading to grant of licences to applicants who were not eligible. Eighty-five out of the 122 licences issued in 2008 were found to be issued to companies which did not satisfy the basic eligibility conditions set by the DoT and had suppressed facts, disclosed incomplete information and submitted fictitious documents forgetting UAS licences and thereby access to spectrum.[30]

Presumptive Value of Spectrum Allocated to 122 New UAS Licensees and 35 Dual Technology Licensees in 2007–08

Any loss ascertained while attempting to value the 2G spectrum allocated to 122 licensees in 2008 can only be 'presumptive', given the fact that there are varied determinants like its scarcity value, the nature of competition, business plans envisaged, number of

---

[29]https://www.thehindubusinessline.com/todays-paper/Executive-summary-of-CAG-report/article20024712.ece; Para 4.6, http://www.thehindu.com/migration_catalog/article15671228.ece/BINARY/Executive%20Summary

[30]https://www.thehindubusinessline.com/todays-paper/Executive-summary-of-CAG-report/article20024712.ece; Para 4.7.1, http://www.thehindu.com/migration_catalog/article15671228.ece/BINARY/Executive%20Summary

operators, growth of sector, etc., which, depending upon the market situation, would throw up the price that it commands at a given point of time. Instead of attempting to come to a specific value of 2G spectrum, which could have been possible only through an efficient market discovery process, we have looked at the various indicators to assess a possible (presumptive) value, from the records made available to Audit rather than going for any mathematical/econometric models.[31]

On 5 November 2007, through a letter addressed to the Hon'ble Prime Minister, STEL Limited, who was a prospective licensee, having applied for UASL in July/September 2007, had offered to pay a higher price in the shape of additional revenue share for next ten years. The offer was enhanced by the firm with a stipulation to further revise it upwards, in case of any counter bid. At the prices offered by the Company, value of 122 new licences and 35 Dual Technology licences after discounting for the receivables in future years works out to ₹65,909 crore as against ₹12,386 crore actually received.[32]

Auction of 3G spectrum was recommended by TRAI in its Report submitted to the Government in September

---

[31] https://www.thehindubusinessline.com/todays-paper/Executive-summary-of-CAG-report/article20024712.ece; Paras 5.1, http://www.thehindu.com/migration_catalog/article15671228.ece/BINARY/Executive%20Summary

[32] https://www.thehindubusinessline.com/todays-paper/Executive-summary-of-CAG-report/article20024712.ece; Para 5.2, http://www.thehindu.com/migration_catalog/article15671228.ece/BINARY/Executive%20Summary

2006. In its Report of 2010, they have observed that it was fair to compare 2G with 3G and recommended 3G prices to be adopted as current price of 2G spectrum in 1800 Mhz band. If these recommendations, which have not been accepted by the Government so far are taken into account, then the value of 2G spectrum allotted to the 122 new licensees and 35 Dual Technology licences would be much higher at about ₹1,52,038 crore as against the amount actually received.[33]

Many of the new UAS licensees of 2008 have been able to attract substantial amount of Foreign Direct Investment (FDI). Value of a new company with no experience in the Telecom sector can primarily be taken as that of the licence and access to spectrum. This would have been the prime consideration for foreign companies while infusing large amount of capital in the form of equity in these companies shortly after award of licence. Based on this indicator, value of a pan India licence works out between ₹7,758 crore and ₹9,100 crore as against ₹1,658 crore priced by DoT. The total value for 122 new licences and 35 Dual Technology licences would be between ₹58,000 to ₹68,000 crore as against the actual revenue of ₹12,386 crore realized.[34]

---

[33]https://www.thehindubusinessline.com/todays-paper/Executive-summary-of-CAG-report/article20024712.ece ; Para 5.3, http://www.thehindu.com/migration_catalog/article15671228.ece/BINARY/Executive%20Summary

[34]https://www.thehindubusinessline.com/todays-paper/Executive-summary-of-CAG-report/article20024712.ece; Para 5.4, http://www.thehindu.com/migration_catalog/article15671228.ece/BINARY/Executive%20Summary

Thus, on the values determined through various indicators, the presumptive value of 2G spectrum on account of grant of 157 licences in different circles during 2007–08 would be in the range of approximately ₹58,000 crore to ₹1,52,038 crore.[35]

Value of Additional Spectrum Allotted to 13 Existing Operators beyond Contracted Quantities

Spectrum was allotted by DoT to the existing operators beyond the contra limits without imposing any upfront charge for such allotment. The value of spectrum held by 13 operators for 51 circles based on the 2001 rates worked out to ₹2,561 crore. Based on the above indicators, value would be in the range of ₹12,000 crore and ₹37,000 crore. The TRAI's recommendation (2010) for charging this additional quantity of spectrum has not been accepted by the Government so far.[36]

Presumptive Loss of Spectrum Allocated to 122 New UAS Licensees and 35 Dual Technology Licences in 2007–08

'If price is calculated at 3G rates...the value works out to ₹1,11,512 crore against ₹9,014 crore realised by the

---

[35] https://www.thehindubusinessline.com/todays-paper/Executive-summary-of-CAG-report/article20024712.ece; Para 5.5, http://www.thehindu.com/migration_catalog/article15671228.ece/BINARY/Executive%20Summary

[36] https://www.thehindubusinessline.com/todays-paper/Executive-summary-of-CAG-report/article20024712.ece; Para 4.10, 5.5, http://www.thehindu.com/migration_catalog/article15671228.ece/BINARY/Executive%20Summary

government,' the auditor had noted. Combined with the price discovery for the dual technology and beyond contracted quantity of 6.2 MHz, the total price fetched could have been in excess of ₹1.76 lakh crore.

The first port of call for the CAG Report was the Public Accounts Committee (PAC) of Parliament where several eminent members other than Congress party MPs expressed serious concerns about the concept of 'presumptive loss'. The CAG of India was summoned to explain it and was cross-examined extensively. Ultimately, it seems that Murli Manohar Joshi prevailed on the Opposition members to go along in the interest of the Opposition.

Beyond the CAG Report

In early November 2010, (Late) J. Jayalalithaa accused the then Tamil Nadu Chief Minister M. Karunanidhi of protecting Raja from corruption charges, calling for Raja's resignation. By mid-November, Raja resigned. At that time, Vinod Rai issued show-cause notices to Unitech, S Tel, Loop Mobile, Datacom (Videocon) and Etisalat to respond to his assertion that the 85 licences granted to these companies did not have the capital required at the time of making application or were otherwise disqualified.

In June 2011, Prime Minister Dr Manmohan Singh, during an interaction with editors, criticized the CAG for commenting on policy issues. He suggested, 'it should limit the office to the role as defined in the Constitution.'[37] After

---

[37] https://www.ndtv.com/business/cags-role-to-ensure-optimal-implementation-of-policy-says-chief-government-auditor-303183

the PM's public criticism, the CAG purportedly conducted a 'rigorous internal appraisal and stood by its findings, citing additional events as corroboration. The CAG reiterated that there was 'an undeniable loss to the exchequer', the calculation of which was based on three estimates: the 3G auctions and the Swan and Unitech transactions of subsequent sale of equity. It cited the Supreme Court ruling of 2 February 2012 that the actions of Raja and officers at the DoT were 'wholly arbitrary, capricious and contrary to the public interest, apart from being violative of the doctrine of equality. The material produced for the quote showed that the MoC&IT wanted to favour some companies at the cost of the public exchequer.' It said its estimate of loss of ₹1.76 lakh crore was justified, since the May 2012 TRAI collation of reserve prices for 2G spectrum was about the same as that in the November 2010 CAG Report. The TRAI had recommended a reserve price for 2G spectrum of ₹180 billion for a pan-India 5 MHz licence, higher than the 3G value of ₹167.50 billion for 5 MHz used by the CAG for arriving at a loss figure of ₹1,760 billion. It concluded that it was only examining the 'implementation of policy', and that policy-making was Government's prerogative. Yet in effect, CAG ended up questioning the policy itself.

4
FACTS FROM JUSTICE SHIVRAJ V. PATIL'S REPORT

In order to put the issue in correct perspective, the UPA government decided that the background of the fastest growing sector had to be better understood. Much had been made of decisions without an objective knowledge of facts that were intrinsic to the growth of the sector. At that stage, it was not entirely clear what was the way forward and how protracted the matter would be. The Supreme Court was passing orders in quick succession and each time there was fresh crisis to deal with. In every inquiry into the matter, it was important to look at the past to justify the present to the extent possible. On the other hand, people who were wielding the sword of accusation tried to steer clear of any legacy issue and keep the matter simple.

The One Man Committee (OMC) was set up on 13 December 2010 to probe and place in the public domain, the genesis of the policy of FCFS in the grant of UASL as well as grant of BSL licences, and spectrum allotment and decisions taken from time to time since 2001–09 by successive governments. The entire report submitted on 31 January 2011 was shared with investigation agencies to determine culpability of all public servants involved in the grant of licence/spectrum between 2001 and 2009.

The OMC report examines the appropriateness of procedures followed by DoT in issuance of licences and allocation of spectrum during the aforementioned period. The report discusses the beginning and growth of the telecom sector in India.

In the *first term of reference*, it studies the circumstances and developments in the telecom sector that led to the formulation of the NTP 1999. According to the report, it was not until the Sixth Five Year Plan that the telecom sector caught the attention of planners, and in the Eighth Plan (1992–1997), the allocation of funds for the telecom sector was also increased.

The initial phases of telecom reforms started in 1984 when the Centre for Development of Telematics (C-DOT) was set up for developing indigenous technologies, and permissions were given to the private sector to manufacture subscriber equipment. Following this, MTNL and Videsh Sanchar Nigam Ltd. (VSNL) were set up in 1986.[38] Thereafter, in 1989, the Telecommunications Commission was established.[39]

---

[38]See Justice Shivraj V. Patil's Report on the 'Examination of Appropriateness of Procedures followed by the DoT in Issuance of Licences and Allocation of Spectrum during the period 2001–2009', p. 2.
[39]Ibid.

Policymakers realized that the telecom sector was one of the fastest growing sectors with a huge impact on the economy. In view of the above, the government formulated the NTP 1994. Some of the important objectives of the 1994 policy included affording telecommunication for all, providing certain basic telecom services at affordable and reasonable prices to everyone, giving world-standard telecom services, creating a major manufacturing base and protecting the defence and security interests of the country. Undoubtedly, it was the first effective step towards deregulation, liberalization and private sector participation.

It is interesting to note that the main reason for the departure from NTP 1994 and formulation of NTP 1999 was that the actual revenues realized from the projects were far lesser than the projections rendering operators incapable of arranging finances for their projects. Since the targets could not be achieved, the government recognized the need to take a fresh look at the policy framework of the telecom sector. Further, some of the objectives of NTP 1994 had remained unfulfilled due to advancements in the sector. Thus in order to meet India's vision of becoming an IT superpower and develop a world-class telecom infrastructure, the government formulated another NTP which came to be effective from 1 April 1999.

Marking its tremendous significance, the objectives of NTP 1999 were to make available affordable modes of communication for citizens, provide universal service to all uncovered areas, encourage development of telecommunication in remote, hilly and tribal areas of the country, protect defence and security interests, enable telecom companies to become truly global players and achieve efficiency and transparency in spectrum management.

Further, the NTP 1999 stipulated that the government would invariably seek the TRAI's recommendations on the number and timing of new licences before taking a decision on the issue of new licences in future.[40] Also, as per the policy, spectrum utilization could be reviewed from time to time keeping in view the emerging scenario of spectrum availability, optimal use of spectrum, requirements of the market, competition and other interests of the public.[41]

It was an accepted fact that the proliferation of new technologies and the growing demand for telecommunication services had led to a manifold increase in the demand on spectrum. Consequently, it is essential that the spectrum was utilized efficiently, economically, rationally and optimally. Transparency was required in the process of allocation of frequency spectrum for use by a service provider and the same had to be examined in the light of ITU guidelines.[42] According to NTP 1999, based on recommendations of TRAI, a fourth group of cellular operators was introduced by following a multistage bidding process in the year 2001.

In the *second term of reference*, the report studies the internal procedures adopted by DoT during 2001 to 2009 in order to enable a better understanding of the internal procedures, the organizational structure and business/functioning of different wings. Hierarchy has been examined while mentioning the key policy perspectives of the Ninth

---

[40]Ibid., p. 11.
[41]Ibid., p. 13.
[42]Ibid., p. 17.

and Tenth Five Year Plans.

Also, the procedure adopted for the grant of basic service licence during 2001–03, CMTS licences during 2001–03, UAS licences during 2004–07 and 2008–09, and allotment of spectrum to CMTS, BTS and UASL have been deduced separately.

Thus, it has been gathered that the Telecom Commission is responsible for the following:

a) Formulating policy of the DoT for approval of the government;
b) Preparing the budget for the DoT for each financial year and getting it approved by the government; and
c) Implementation of government's policy in all matters concerning telecommunications.

The Wireless Planning and Co-ordination (WPC) wing in the DoT deals with the policy of spectrum management, wireless licencing and frequency assignment. Spectrum Allocation Policy is contained in National Frequency Allocation Plan (NFAP) to be drawn periodically, which is based on the International Radio Regulations framed and revised from time to time by ITU.[43]

The report dwells on the procedure adopted for grant of basic service licences during 2001–03. Some of the terms worth mentioning are as follows:

a) That the BSLs could be granted without any restriction on the number of operators.
b) That the applicant has to satisfy the eligibility criteria as

---

[43]NFAP (1981) has been revised in the years 2000, 2002 and 2008.

to the minimum paid up capital, combined net worth of promoters, experience in telecom sector and foreign equity, etc.

c) That the applicant has to submit the application in a prescribed form with stipulated documents like business plan along with its funding arrangement for financing the project, detailed rollout plan, etc.

Interestingly, for the grant of CMTS licences during 2001–03, the DoT sought recommendations of the TRAI on the appropriate level of entry fee, percentage of revenue to be shared with the licensor, definition of revenue for the purpose and the basis of selection of new operators and any other issue considered relevant in the light of NTP 1999.[44] The TRAI forwarded its recommendations dated 23 June 2000, about the basis of selection of new operators through a competitive process by multistage bidding from amongst applicants meeting predetermined eligibility criteria and the entry fee to be determined on the basis of the highest bid.

Also, the report discusses in detail the procedure for grant of UASL which was granted on the FCFS basis, apart from all other conditions required to fulfil the stipulated eligibility criteria. It is important to note with due caution that further changes were introduced by the DoT during 2008–09 for the grant of UASLs. The important change relating to spectrum policy was that the entire spectrum, excluding spectrum in 800, 900 and 1,800 bands, had to be auctioned in future so as to ensure efficient utilization of

---

[44]By the communication No. 842-153/99—VAS (Vol. IV) dated 23-4-1999.

this scarce resource.[45]

On 19 October 2007, the DoT issued a press release[46] notifying that the recommendations of TRAI to have no cap on the number of excess providers in any service area has been considered and accepted by the government and that the allocation of spectrum and grant of wireless licence shall be subject to availability. In case UASL is not allocated spectrum due to non-availability, the licensee shall endeavour to roll out services using wireless technology.

On 2 November 2007, A. Raja took the view that the opinion of the Minister for Law and Justice was out of context and directed existing policy for the grant of new UASLs followed till then, i.e. the FCFS basis be continued.[47] Accordingly, he informed the then prime minister Dr Manmohan Singh on various issues like criteria for allotment of UASL, use of dual spectrum, issue of new licences and LoIs, etc. In the said letter, it was mentioned that FCFS policy for granting LoI was followed by the DoT for grant of UASL, which meant an application received first would be processed first, and if found eligible, would be granted LoI. It was further mentioned that FCFS is also applicable for grant of licence and compliance with LoI conditions and therefore, any applicant who complied with the conditions of LoI first, was to be granted UASL first.

The report examines in detail that on 10 January 2008, the DoT issued a press note on its website and on the

---

[45]Ibid., p. 57.
[46]http://www.dot.gov.in/press-release/press-releases
[47]See Patil's Report on the 'Examination of Appropriateness of Procedures followed by the DoT', p. 61.

website of Public Information Bureau (PIB) at 1.47 p.m. notifying that the former had decided to issue LoI to all applicants eligible on the date of application, who applied by 25 September 2007.[48] It also notified that the DoT had been implementing a policy of FCFS for the grant of UASL.

On the same day, i.e. 10 January 2008, a second press release was issued by DoT on its website at about 2.45 p.m. to the following effect:[49]

> Sub: UASL applicants to depute their authorized representative to collect responses of DoT on 10.1.2008.
>
> The applicant companies who have submitted applications to DoT for grant of UASL in various service areas on or before 25.9.2007 are requested to depute their Authorised signatory/Company Secretary/ authorized representative with authority letter to collect response(s) of DoT. They are requested to bring the company's rubber stamp for receiving these documents to collect letters from DoT in response to their UASL applications. Only one representative of the Company/group Company will be allowed. Similarly, the companies who have applied for usage of dual technology spectrum are also requested to collect the DoT's response. All above are requested to assemble at 3:30 p.m. on 10.1.2008 at Committee Room, 2nd Floor, Sanchar Bhawan, New Delhi. The companies which fail to report before 4.30 p.m. on 10.1.2008,

---

[48]Ibid., p. 65.
[49]Ibid.

Facts from Justice Shivraj V. Patil's Report • 73

the responses of DoT will be dispatched by post. All eligible LoI holders for UASL may submit compliance to DoT to the term of LoIs within the prescribed period during the office hours i.e. 9.00 a.m. to 5.30 p.m. on working days.

On 10 January 2008, between 3.30 p.m. and 4.30 p.m., LoIs were issued to eligible applicants in Committee Room, 2nd floor, Sanchar Bhawan, New Delhi, where the office of DoT is located. On the same day, four special counters were set up in the reception area at the ground floor of Sanchar Bhawan for receiving LoIs between 3.30 p.m. to 5.30 p.m. The officers of the DoT received the LoIs and recorded the time of receipt by referring to a digital clock mounted on the wall.

The *third term of reference* of the said report examines as to whether the procedures adopted by DoT during the period 2001–09 for the issue of telecom access service licences and allocation of spectrum to all telecom access services licensees were in accordance with extant policies and directions of the DoT/government.

Further, the report mentions that Article 14, 19(1) (g) and 21 mandate that any procedure adopted for issuance of licences should be fair, transparent and in public interest, and that the selection criteria must be certain and free from any ambiguity. Generally, public property owned by the state should be sold by public auction or by inviting tenders in order to secure the highest price and also to ensure fairness in the activities of the state and public authorities.[50] There

---
[50]Ibid., p. 83.

may be situations necessitating departure from the rule in public interest, but then such instances must be justified by compulsions and not by compromise.[51]

The FCFS procedure adopted and applied was clearly inconsistent and was without any nexus with the objective of the selection of UAS licensees pursuant to NTP 1999.[52] The report states that by applying FCFS, the best eligible applicants' offer could stand excluded. This was opposed to the principles of level playing field amongst prospective applicants. The criterion of FCFS as adopted by the DoT was neither contemplated nor was it consistent with the NTP 1999, the recommendations of TRAI and the cabinet decision. Added to this, the basis of reckoning to apply FCFS was not consistently followed. Thus, the report inferred that the internal procedures adopted by the DoT have not been in tune with the extant policies and the directions of DoT/Government.

The report in its *fourth term of reference* examines as to whether the procedures were followed consistently and also identifies specific instances of deviation from and inappropriate application of the laid down procedures.

As per the guidelines for the grant of BSLs dated 25 January 2001, there was no provision for extending time to rectify deficiencies in the applications. The guidelines stipulated that the application, if deficient, will be rejected.[53]

---

[51]Ibid.
[52]Ibid., p. 87.
[53]Under ToR-2, the procedures adopted by the DoT during the period 2001–09 for issue of telecom access service licences as also allocation of spectrum to all telecom access service licensees, have been examined.

However, on 16 February 2001, a decision to grant time to applicants for BSLs without issuing any formal order and notifying the same to all applicants was taken arbitrarily.[54] Such a decision in effect overrode the notified guidelines without giving any clear indication as to the specific cases in which extension could be granted. Further, as per the notified guidelines for grant of BSLs dated 25 January 2001, it was stipulated that in case an applicant was found eligible for grant of BSL, the applicant would be required to deposit entry fee and submit bank guarantees and sign the licence agreement within thirty days, failing which the offer of grant of licence was to stand withdrawn at the expiry of the permitted period.

The guidelines had no provision for extension of time for compliance with LoI. However, extension of time spreading to several months was granted to Tata Teleservices Ltd. for compliance with LoIs issued to it in respect of Maharashtra, Haryana, Kerala, Punjab and Rajasthan service areas. This was contrary to the laid down procedure. Stipulations regarding compliance within the prescribed time were rendered meaningless.

Also, time was extended for rectifying discrepancies in the application of Idea Cellular Ltd. dated 4 August 2005 for grant of UASLs for the Mumbai service area without the approval of the Member (F).[55] While time could be extended

---

[54]See Patil's Report on the 'Examination of Appropriateness of Procedures followed by the DoT', p. 90.
[55]For granting UASLs, though it was decided to adopt the procedure applicable for grant of BSLs, the procedure as per decision in File No. 10-1/2001-BA-II, 2/C dated 16-02-2001, as applied to BSLs for extension

by almost 30 days, time to rectify deficiencies was extended for over a year for Idea Cellular Ltd[56] and LoI was issued only on 20 November 2006. Thus, applying FCFS for the grant of UASL extension of time for over a year without the concurrence of M (F) amounted to contravention of the laid down procedure.

Also, in case of Bharti Airtel Ltd., an allotment of 2 + 2 MHz beyond 8 + 8 MHz of spectrum was made for Delhi service area on 17 July 2003, though no criteria for allotment beyond 8 + 8 MHz existed. The allotment seems to have been made in anticipation of the report of Lalwani Committee, which was approved by MoC&IT Arun Shourie on 18 August 2003. In the absence of the laid down procedure much less published/announced one, allotment of spectrum (additional) beyond 8 + 8 MHz was improper.[57]

Thus, it is noted that there appears to have been contravention, deviation, inappropriate application and violation of the underlying principles of the laid down procedures arbitrarily from 2001 to 2009.

Under the *fifth term of reference* it was examined as to whether the procedures adopted were fair and transparent and in accordance with the principles of natural justice. In examining the procedures adopted for issuing telecom access licences and allotment of spectrum, the concept of fairness and transparency, keeping with the principles of

---

of time for rectifying deficiency in application with the prior approval of Member (Production) and Member (Finance) was not followed.
[56]File No. 10-1/2001-BA-II, 2/C dated 16-02-2001
[57]See Patil's Report on the 'Examination of Appropriateness of Procedures followed by the DoT', p. 103.

natural justice, would import the following:

i) They should be duly notified to public so that all the interested and intending applicants had knowledge of the same.
ii) They ought to be rational, objective, relevant and certain.
iii) They ought to be notified in advance giving reasonable time to all to participate with equal opportunity.
iv) All applicants ought to be given equal opportunity and equal treatment while applying such procedures. In case of any relaxation, wherein prescribed/ adopted procedure is allowed for valid reasons, such relaxation should apply to all uniformly and should not result in any discrimination.
v) They should be consistent with the policy and statutory provisions to ensure fairness and transparency.

Keeping the above in view and looking at the procedures adopted as already examined in the *second term of reference* for grant of access service licences and allocation of spectrum, following are the observations:

a) In the year 2001, licences for CMTS (to four operators) were issued by adopting a multistage bidding process after due publicity. The procedure adopted was fair and transparent. Grant of BSLs in the year 2001 upto the advent of UASL regime was in accordance with guidelines for grant of BSLs dated 25 January 2001. Clause 23 of the said guidelines prescribed that BSL operators who required spectrum for offering WLL would be granted the same on the FCFS basis. While said guidelines provided for FCFS basis, the exact point or event for reckoning priority amongst various applicants (for instance, the date of application for grant of access licence, the date of grant for access licence, the date of compliance with LoI, the date of application for allotment of spectrum or any other date) were not specified, thus leaving room for subjectivity and arbitrariness. In the absence of certainty, FCFS criteria were opposed to principles of fairness and transparency.

b) Though initially considering that all Basic Service Licensees did not require allotment of spectrum and spectrum was available for those who required the same (for WLL), the FCFS criteria for allotment of spectrum even without any indication as to the exact point or event for reckoning priority may not have presented any difficulty. Subsequently, on 24 November 2003, a decision having been taken by the DoT for applying FCFS basis as criteria for grant of access licence, no guidelines were drawn up and no procedure was formulated specifying the point or event for reckoning the priority amongst applicants, such as the date of application or the date and time of LoI or the date of compliance with LoI. Considering scarcity of spectrum and a large number of applications that were made for the grant of UASL, adopting FCFS as the criteria for issuing licences for UASL without having fixed the point/event for determining priority, made it wholly subjective and arbitrary as the decision was left to absolute discretion of the authority to change the point of reckoning. This was also opposed

to the principle of fairness and transparency.
c) At the time of the decision taken in the year 2003 to consider the applications for the grant of UASL on FCFS basis, applications from two operators, namely, Bharti Airtel Ltd. and Tata Telecom Services Ltd. were already pending. As such, FCFS as the basis for grant of UASL could have been unfair to other intending applicants who could not anticipate such decision and consequently, could not make applications to take advantage of this policy. Besides, the FCFS criterion in its application to the grant of access licences lacks objectivity and is opposed to the principles of a level playing field.
d) In the application form for BSLs in terms of guidelines dated 25 January 2001, while requirement of paid up equity capital and combined net worth of the promoters was stipulated, the guidelines did not prescribe documents required to be submitted by an applicant to establish the same. Similarly, while guidelines for the grant of UASL dated 14 December 2005 also contained stipulations as to minimum combined net worth, the documents required to establish the same was not specified. This had the potential to cause delay in processing on account of the likelihood of questions being raised at different stages for ascertaining the correctness of information furnished. This lack of stipulation made the procedure open to abuse at the processing stage. Hence it was not fair.
e) In guidelines dated 25 January 2001 for issue of BSL as also in the guidelines announced on 5 January 2001 for issue of licences of CMTS to fourth operator, eligibility

criteria included experience in telecom sector. However, the nature and extent of experience were not specified, thus taking away the objectivity in assessment. This ambiguity could lead to inconsistencies in dealing with applications.

f) While guidelines for issue of BSL/application for UASL require submission of business plan along with its funding arrangements for financing the project, no specific requirement as to contents of the business plan or extent/source/commitment of funding arrangement was stipulated. This also introduced an element of uncertainty and subjectivity as it was left to the discretion of officers to decide what would qualify as 'business plan along with its funding arrangements for financing the project.'

g) While the guidelines dated 14 December 2005 stipulate the restriction on substantial equity of applicant in another operator in the same service area, the term equity, particularly whether the same includes preferential shares, was not defined resulting in uncertainty/subjectivity.

h) In reference to 'Note of Chairman, Telecom Commission dated 16 February 2001' in File No.10-l/2001-BS-II, 2/C, a decision was taken to extend the time for rectifying deficiency in the application for the grant of BSL within a reasonable period of almost 30 days. However, the decision was neither incorporated in the guidelines dated 25 January 2001 nor published for all intending applicants. Hence, it was non-transparent and unfair.

i) As per the guidelines for grant of BSL after the issuance of LoI, its terms were required to be complied within three months. There was no provision for extension of period of compliance or any laid down guidelines for such compliance. Such being the case, in the absence of guidelines, extension of time for compliance with terms of LoI was granted to some of the applicants for different reasons and for varying periods on several occasions. Similarly, in the case of guidelines for UASLs, no specific provision existed for extension of time for compliance with terms of LoI. However, there were instances when time for compliance was extended on a case to case basis. Lack of uniformity in guidelines/ principles governing extension of time for compliance with terms of LoI was opposed to principles of fairness and transparency.

j) Press Note 5/2005 issued by the Department of Industry, Policy and Promotion, Ministry of Commerce and Industries, while enhancing the FDI ceiling from 49 per cent to 74 per cent, had stipulated that infusion of FDI above 49 per cent up to 74 per cent shall require approval by FIPB and compliance with certain conditions. By virtue of Press Note 3/2007, the earlier guidelines were superseded. Subsequent to issuance of Press Note 3/2007[58], though the DoT seemed to have dispensed with the requirement of compliance with the conditions stipulated in Press Note 5/2005, the guidelines dated 14 December 2005 were not amended to reflect

---

[58]See Patil's Report on the 'Examination of Appropriateness of Procedures followed by the DoT', p. 110.

such a decision, thus leaving scope for delay in processing on account of individual understanding and assessment of the concerned officers as to the implication of Press Note 3/2007.

k) After the decision of the Minister to grant UASLs to all applicants who had applied on or before 25 September 2007 and to recur priority from the date of compliance with LoI, the first press note dated 10 January 2008 was issued at about 1.47 p.m. notifying the same. It was further notified that the DoT has been implementing FCFS basis for grant of UASLs under which initially an application which is received first will be processed first and thereafter, if found eligible, applicant will be granted LoI, and then whosoever complies with the conditions of LoI first, will be granted UASL. The said press note though for the first time had notified the decision of the DoT to accord priority to applicants who complied with LoI first, wrongly mentioned that the DoT had been implementing such a policy in the past. Further, the first press note dated 10 January 2008 was published on the websites of the DoT and PIB only. This first note contained critical information as to drastic change in procedure followed by the DoT hitherto as the priority already acquired by applicants by virtue of date of submission of applications was to change. This note affected the rights of applicants inter se. The publication on the websites without publications in newspapers and without individual communications to all the applicants was opposed to the requirements of transparency and fairness.

l) On the same day, i.e. 10 January 2008, a second press release was issued by the DoT at about 2.45 p.m., i.e. in less than an hour's time, requiring the representatives of applicants to collect LoIs on the same day between 3.30 p.m. and 4.30 p.m. The second press release requiring the applicants to collect LoIs simultaneously, in effect, took away the priority acquired by applicants who had applied earlier. This release was very significant as it was only after collecting the LoI that an applicant could comply and the applicant complying earlier was to get priority. This press release was again published on the websites of DoT and PIB. No record of individual communications to all applicants having been sent is available. Thus, it was unfair and non-transparent.

m) The LoI for grant of UASL issued on 10 January 2008 stipulated fifteen days as the period within which the terms of LoI had to be followed by an applicant. Having stipulated the period for compliance, there was no justification in granting priority to an applicant who complied with LoI earlier than fifteen days and also earlier than other applicants comparatively, rendering it unfair.

n) In case of some of the applicants, it was noticed that while pursuant to the second press note dated 10 January 2008, they had collected LoIs on the said date and had also submitted compliance on the same day, the demand draft for payment of entry fee was dated earlier to 10 January 2008. In the case of Unitech Infrastructure Pvt. Ltd., Volga Properties Pvt. Ltd., Azka Projects Ltd., Azare Properties Ltd, Unitech Builders & Estates Pvt. Ltd.,

Adonis Projects Pvt. Ltd., Hudson Properties Ltd., while the LoIs were issued on 10 January 2008 and compliance was submitted on the same day, the demand drafts for payment of entry fee were dated 24 December 2007. In case of Idea Cellular Ltd., while LoI was issued on 10 January 2008 and the compliance was submitted on the same day, the demand draft for payment of entry fee was dated 8 January 2008. For the first time, the procedure for according priority to those applicants who complied with the terms of LoI first, was mentioned in the letter of MoC&IT to the prime minister dated 26 December 2007. The decision in the files of the DoT to the said effect was taken only on 7 January 2008. The decision was made public through a press release dated 10 January 2008. The submission of the demand draft dated earlier than the notification of the decision indicates the possibility of some of the applicants having known the changed procedure to be notified in advance and in contemplation of such change had already kept ready the demand drafts for payment of entry fee. This also was opposed to the principles of fairness and transparency whereby some of the applicants had the benefit of receiving in advance the information for change in the procedure thus gaining an edge over other applicants. In the matter of grant of spectrum, in the absence of information as to its availability and no time frame having been fixed for allotment of spectrum, the procedure tended to be unfair, arbitrary and selective. There could be delay also on this count.

Having taken note of the requisites of fairness and transparency, the specific instances suggested the lack of the same in the procedure adopted by the DoT in granting access licences and allotment of spectrum during 2001–09.

Under the *sixth and seventh terms of reference*[59], the shortcomings and lapses in the implementation of the laid down procedures have been revealed and the identity of the public officials who appear to be prima facie responsible, have been provided by name and others, identified through reference to their designations/responsibilities, for want of information in the short time available. The report did not propose criminal culpability in view of being unable to issue notices to individuals concerned and suggested that departmental disciplinary action with due process.

Under the *eighth term of reference*[60], the remedial measures to avoid deficiencies in formulation of procedures in future and lapses in implementation of the laid down procedures have been suggested concretely. It has been advised that a selection procedure which is reasonable, fair, transparent and based on merit be devised instead of FCFS. It has been further recommended that procedures which specify a time frame for receiving/scrutinizing applications to be formulated, which further intimate the eligible/ineligible applications, process them and impart the decision to the applicant formally in writing.[61]

Procedures formulated based on policy guidelines/

---

[59]See Patil's Report on the 'Examination of Appropriateness of Procedures followed by the DoT', pp. 124–35.
[60]Ibid., pp. 136–43.
[61]Ibid., pp. 136–43; Para 8.1 (ii) and (iii), p. 137.

directions of the government should be approved and authenticated by the Telecom Commission. In case of change in procedures, the same must be notified well in advance before implementation. The procedures formulated must not only spell out criteria, but also specify documents required to satisfy the eligibility. A comprehensive checklist should be drawn up based on the procedures and must be included in the prescribed application form itself.

The discontinuation of practice of 'Internal Telecom Commission' has also been suggested. All important matters must be placed before 'Full Telecom Commission'. A need for drawing up a procedure to make it mandatory for placing recommendations of the TRAI before the Telecom Commission within a specified time frame has also been suggested. Drawing up of a self-contained office memorandum indicating procedures to be followed relating to grant of Access Service Licences and allotment of spectrum, with approval from the Telecom Commission has been advised. The DoT should put spectrum allocations to various agencies in the public domain. All spectrum allocations should be audited to determine efficient and proper utilization of the allotted spectrum.

Comprehensive spectrum reforms have been recommended for efficient utilization. A need to incentivize vacation of unutilized spectrum and a penalty on hoarding has been suggested. The process of allotment of spectrum has been advised to be delinked from access licences and the entry fee/spectrum pricing needs to be structured accordingly. Auctioning of spectrum by formulating suitable design has been suggested. The need for a comprehensive new legislation

'Radio Communications Act' has also been felt.[62]

The OMC observations suggest, 'In suo moto recommendations dated 27 October 2003, TRAI had recommended that additional players could be introduced through multistage bidding process, which was also accepted by the Union Cabinet on 31 October 2003. However, in deviation with the said requirements, the Secretary DoT, on the contrary, on 17 November 2003, approved formulation of procedure of accepting the applications for grant of UASLs by adopting a procedure similar to the procedure adopted for grant of BSL.'

Further, on 24 November 2003, the minister approved the formulation of the procedure for grant of UASLs on the basis of FCFS as against through multistage bidding process. All this was clearly in deviation from extant policies. The DoT, contrary to the said recommendations, 'formulated the procedure on 24 November 2003 to collect entry fee from new operators at the rate paid by fourth group of operators thus deviating from the policy framework of NTP 1999'.[63]

It appears strange that such a comprehensive report was not heard of once it was submitted to the government during the tenure of Kapil Sibal, Raja's successor. But, of course, one thing is clear that instead of concluding that the allocation under Raja was simply a continuation of the procedure followed by his predecessors in the NDA government, something that the trial court ultimately held, the OMC found fault with FCFS itself, leave alone its

---

[62]Ibid., p. 142.
[63]Ibid.

distortion. Perhaps this is the reason why the government did not find any use for the report and instead fell back on the JPC. However, similar to the trial court the OMC also found fault with the manner in which officials concerned sought to deal with their responsibilities. This leads to the constant debate about apportioning responsibility in our system between the civil servants and political executive. However, the time-tested convention of the political arm taking responsibility before Parliament and providing a shield to their civil servants is clearly under stress, as the system hounds political leaders and justifiably so, or imposes criminal liability upon them. In the process, it is not surprising that officials attempt to pass the buck.

5
COLOURS OF THE RAINBOW: THE JPC REPORT

When the Indian media began to cite the CAG report identifying the loss at ₹1.76 lakh crore, the opposition parties unanimously demanded the formation of a Joint Parliamentary Committee (JPC) to investigate the matter. The government stoutly rejected their demand. The Opposition again pressed for a JPC when the winter session of Parliament began on 9 November 2010. Again, their demand was rejected. The demand for a JPC gained further momentum when the CAG report was tabled in Parliament on 16 November 2010. The Opposition blocked the proceedings, again pressing for a JPC; the government again rejected their demand, creating an impasse. The Speaker of the Lok Sabha Meira Kumar unsuccessfully attempted to resolve it. The winter session of Parliament concluded on 13 December 2010. Although 22 new bills were planned to be introduced, 23 pending bills passed and three bills withdrawn, Parliament functioned for only nine hours.

On 22 February 2011, after resisting demands from the Opposition for over three months, the government finally announced that it would form a JPC.

Half the members of the JPC consisted of UPA members and the other half included members of the Opposition. Twelve were from the Lok Sabha, and eight from the Rajya Sabha. Of the Lok Sabha MPs, eight were from the Congress party and four from the BJP.

Although the JPC, under the Chairmanship of MP P.C. Chacko, considered several aspects of the mobile telephony sector, the critical dimension was the assessment of the financial impact on account of the issue of licences and spectrum inter alia under minister Raja. The summary of the JPC Report submitted on 25 October 2013, relating inter alia to examination of matters pertaining to allocation and pricing of telecom licences and spectrum is as follows:

Summary of the Report

The First Chapter discusses reports relating to allegations of huge losses in 2008–10. It takes note of the complaints regarding the FCFS policy, looked into by the Central Vigilance Commission (CVC) in respect of allotment by the DoT, and the CBI investigation of the matter is based on the report sent by the CVC.

On 8 November 2010, the Performance Audit Report was submitted by the CAG on allocation of 2G spectrum which projected a loss ranging from ₹57,666 crore to ₹1,76,645 crore.

The Committee held fifty-seven sittings in total. It received briefings and oral evidence by the representatives of DoT and sought extension of time on five occasions for presentations. The Parliament library has kept a track of the Committee record prepared and laid on the Table of the House.[64]

The Second Chapter focusses on the era of the NTP 1994. With the aim to spur deregulation, providing telephone on demand, world-class services at reasonable prices, universal availability to all villages, the government had announced the first NTP on 13 May 1994. To achieve the required targets and to bridge the large resource gap, investment and involvement of the private sector was required. NTP 1994 required companies to maintain a balance in their coverage between urban and rural areas and also be allowed to participate in the expansion of the telecommunication network.[65]

Private sector participation was invited in a phased manner for granting licences to bridge the resource gap. There were two phases. The first phase was held in November 1994 and the second from December 1995 to April 1998. In the duopoly regime of first and second CMTS licences, the right to operate the services in the area was designated through a public authority.[66] In the second phase, auction

---

[64]'Joint Parliamentary Committee (JPC) Report: To Examine Matters Relating to Allocation and Pricing of Telecom Licences and Spectrum', October 2013, http://www.prsindia.org/uploads/media/JPC%20Report%20on%202G/ JPC%20Report%20on%202G.pdf, p. 2.

[65]Ibid., p. 3.ii

[66]Ibid., p. 4.

was conducted. During 1997–98, the six licensees for BTS who, by auction, were chosen to sign the licence agreement defaulted in payment of licence fee. With mounting overdue payments and in July 1997 Cellular Operators Association of India (COAI) seeking concessions on the ground of inadequate revenue, the financial health of the cellular industry was required to be studied by an expert agency and appropriate measures were to be suggested on the basis of its outcome. It was also noted that TRAI had an advisory role under the TRAI Act and would be suitable for the exercise. This was a decision taken at the level of the prime minister on 1 November 1997 but on the understanding that since the TRAI would take a long time, the Bureau of Industrial Cost & Pricing (BICP) be asked to take up the task. Before they could complete it, an informal assessment was sought from ICICI. The Committee was surprised that the then Secretary, A.V. Gokak took that decision without reverting to the PMO.

In May 1998, the DoT confirmed that a report prepared by the private consultant, ICICI, was submitted informally. There were some issues relating to conflict of interest as ICICI was lender to many service providers. Curiously, ICICI recommended an extension of the licence period from 10 years to 15 years. As per the available information and records, the matter was referred to the legal advisor in the DoT for obtaining internal legal opinion but there is no record of obtaining the Attorney General's (AG's) Opinion.[67] The matter was then placed before the Chairman of the

---

[67]Ibid., p. 8.

Telecom Commission and Cabinet approval was sought. At that moment, 11 out of 14 circle operators were in default. On 9 November 1998, the BICP[68] finally submitted its techno-economic report to DoT. It found that the demand in metros exceeded estimates but in the circles it was but 15–45 per cent of the 1996–98 estimates. However, the Bureau viewed that as temporary phenomenon and that eventually, the demand would be robust. The Committee concluded that the government had acted in haste and without prudence in accepting the suggestions of ICICI.

The result of privatization had not been satisfactory even four years after NTP 1994 was introduced. To develop world-class telecom infrastructure in the country, the government required a new telecom policy framework which would facilitate India's vision of becoming an IT superpower. On 15 December 1998, a comprehensive note was prepared by the DoT seeking the help of the AG on the financial problems faced by the licensees. On 20 December 1998, outstanding dues were ₹3,100 crore.[69]

On 21 December 1998, the finance minister received a letter from the minister of communications[70] about the state of the mobile sector and within four days i.e. 24 December 1998, the finance minister replied sharing his concern over non-recovery that would add to the budget deficit of ₹2,800 crore. Meanwhile, the chairman of the Telecom Commission forwarded the note to the AG. On 6 January 1999, the

---

[68]Ibid., p. 11.
[69]Ibid., p. 13.
[70]Ibid., p. 15.

government received advice from the AG.[71] He opined that since the extension had already been granted in principle, the indulgence should be on certain conditions of financial safeguards for the government including some payment (say 20 per cent) to show bona fides. On 9 January 1999, the prime minister confirmed that the advice of the AG was to be followed. As a pro tem measure pending finalization of a new Telecom Policy, the government agreed to extend the date of payment up to 28 February 1999.

The outstanding dues of licence fee against the Basic and Cellular service operators including interest amounted approximately to ₹3,708.10 crore. Six licences were terminated as they failed to pay even 20 per cent of arrears. But three licences were restored subsequently on the payment of dues.[72]

The Committee noted that the DoT, at no point of time, calculated the loss suffered by the exchequer due to the Migration Package. The Committee has, however, been informed that impact on licence fee and spectrum charges collection due to the Migration Package was to the tune of ₹43,523.92 crore which included the sum of revenue, i.e. ₹1,443.58 crore foregone on account of extension of effective date of licence by six months. The DoT has stated that the Migration Package did not provide for recovery of the amount of ₹42,080.34 crore (after excluding ₹1443.58 crore). According to the Migration Package, the telecom licensees were to pay '...one time Entry Fee and Licence Fee

---

[71]Ibid., p. 16.
[72]Ibid., p. 18.

as a percentage share.'

It is thus clear that the telecom policy was gradually evolving with the government responding to emerging challenges. But at several stages there were obstacles that the officials dealing with the policy were woefully unprepared to handle. With the benefit of hindsight many decisions could be faulted but obviously none of that could lead to any criminal liability.

The Third Chapter exhaustively discusses the NTP 1999 and the related Migration Package. The NTP was made effective from 1 April 1999 and was considered and approved by the Cabinet on 26 March 1999. For approving the NTP 1999, the Union Cabinet noted that it is in the interest of the public and it should be applicable uniformly all over the country. It also obtained the advice of the AG to fortify its position.[73] On 6 July 1999, the Cabinet approved the Migration Package.

The Group on Telecom and IT (GoT-IT) submitted a report to prime minister on 26 April 2001 that was accepted the next day. The Group noted that WLL would greatly facilitate the roll out in rural areas at affordable prices. Even for urban areas it was considered ideal to meet congestion and technical impediments. TDSAT dismissed the challenge of COAI on the ground that this pertains to government policy.

On 28 June 1999, the Cabinet Secretary had received a note from the AG for urgent consideration of the matter. On 5 July 1999, the DoT received comments from both the Ministry of Law and the Ministry of Finance. Under the

---

[73]Ibid., p. 19.

Migration Package proposal, certain objections were raised by the Finance Ministry. In the cabinet meeting held on 6 July 1999, the draft note for the cabinet was approved by the prime minister and considering the legal opinion given by the AG and views of financial institutions, the Union Cabinet approved migration of existing licensees to the NTP 1999 regime.[74] The decision taken by the cabinet was a follow-up in regard to NTP 1999. The Package had already been initiated for implementation and operationalization. All CMTS and BTS licensees migrated to the revenue sharing regime for the implementation of the Migration Package. There had been wide coverage regarding the matter in the electronic and print media. The total revenue generated post-migration was around ₹4,144.32 more than the revenue that would have been otherwise generated. However, the government being in caretaker mode because of general elections, the President of India and the Election Commission advised keeping the decision in abeyance. The Delhi High Court too intervened to grant an interim order on 10 August 1999 permitting the Migration Package subject to approval of the Cabinet after the constitution of 13th Lok Sabha.

The TRAI in its report dated 27 October 2000, supported the policy shift from fixed fee to revenue share implemented under NTP 1999, which resulted in a huge growth in the sector. The effective date of licensees of basic and cellular operators that was proposed by the AG was approved by the Cabinet. Of course that came with safeguards on payment of dues.

---

[74]Ibid., p. 21.

The DoT informed the Committee in writing that all the outstanding dues in terms of the migration package as proposed by the AG were recovered from the licensees allowed to migrate to NTP 1999.[75] However, the Committee observed that the government had to forego revenue to the tune of ₹42,080.34 crore due to migration.

The Fourth Chapter discusses the Migration Package in detail. DoT services, termed as BSNL, granted CMTS licences for 21 telecom circles, and Chennai and Calcutta metros on 29 February 2000. After the withdrawal of court cases by CMTS licensees, MTNL provided mobile services in Delhi and Mumbai in terms of the Migration Package on the basis of licences granted.

BSNL and MTNL cellular operators were allotted in the 900 MHz band. The NTP 1999 envisaged the opening up of basic telephone services. The TRAI accessed the process of consultation and sent its recommendations to the DoT on 31 August 2000. The Telecom Commission, after taking the opinion of the TRAI on 9 October 2000, came to the conclusion that the amount of entry fee and spectrum charges as a percentage of revenue should be charged from basic service operators, after considering the recommendations on 21 September 2000.

The TRAI made some policy changes to favour cellular operators for migration due to continuing losses. It also recommended permitting migration to cellular mobile operators based on GSM network infrastructure.[76] On 24

---

[75]Ibid., pp. 25–35.
[76]Ibid., p. 36.

January 2001, the Full Telecom Commission considered and accepted the recommendations of the TRAI.

Accordingly, DoT issued guidelines on 25 January 2001. It also effected consequent changes in the licence for Basic Service Operators. Twenty-five additional BTS licences were issued and as per records, 147 applications were received.

On 27 April 2001, the report submitted by GoT-IT was accepted by the prime minister.[77] The Group desired that steps be taken to stop the violation of licence conditions by BSTs thereby ensuring that the TDSAT judgement should be enforced to restrict BSTs within the Short Distance Charging Area (SDCA) mobility.

On 23 June 2000, the DoT had received the recommendations forwarded by TRAI. These were later approved by the Telecom Commission with some modifications. The fourth cellular operators were allotted spectrum in 1,800 MHz band.[78] On 10 January 2002, the chairman of the Telecom Commission submitted a note to MoC&IT about the grant of additional frequency to the cellular operators facing shortage of spectrum in Delhi and Mumbai. The above proposal was approved by the Minister on 31 January 2002.

The Fifth Chapter discusses the UASL regime in detail. The implementation of NTP 1999 was a consequence of the obliteration of differences between wireless and wireline systems and TRAI's opinion that there was no justification in continuing a service-centric licencing regime.

---

[77]Ibid., p. 39.
[78]Ibid., p. 44.

On 16 January 2003, the TRAI introduced Unified Telecom Licensing in the country. It gave its suo motu recommendations unifying wireless and wireline systems on 27 October 2003.[79] It had a target of 100 million-subscriber strategy. The TRAI requested that the existing system would be replaced by the Unified Licensing Regime/Automatic Authorization Regime in two phases; the basic and cellular services in the immediate first phase, and the second phase were to be followed in six months with fully Unified Regime.

On 4 November 2003, the TRAI sent its opinion on the 'issue of fresh licences for cellular mobile services providers'. Under the chairmanship of the finance minister on 10 September 2003, the GoM's recommendations on telecom matters received its approval from the prime minister and the cabinet considered and approved their recommendations on 31 October 2003.

Guidelines for UASL were prepared and a press release was issued on 11 November 2003. The DoT had pointed out that the recommendations of TRAI in respect to Unified Access Licensing regime were approved by the cabinet.

On 12 November 2003, the DoT received applications in the form used for BSL. The same day, the Chairman of the Telecom Commission approved the acceptance of applications made for grant of UASLs and adopted an application for issue of licences procedure similar to BSL. On 14 November 2003, the then Chairman of TRAI conveyed that the fee for a new Unified Licensee would be the entry fee of the fourth cellular operator. The cellular industry had several

---

[79]Ibid., p. 50.

grievances against the non-implementation of the TDSAT order of August 2003.[80] On 17 November 2003, the DoT requested TRAI to facilitate grant of UASL.

The TRAI initiated the process of framing recommendations on Unified Licensing. It also issued a Consultation Paper on 15 November 2003 and a detailed consultation paper on 13 November 2004. In 2005, the DoT received the recommendations on Unified Licensing Regime sent by TRAI on 13 January of the same year. Its primary objective was to encourage free growth of new application and services in the ICT area. The TRAI's recommendation for Unified Licensing of all telecom services dated 13 April 2007 was helpful for the DoT. On 11 May 2010, it recommended Unified Licence, which excluded broadcasting licences.[81]

The Sixth Chapter discusses the critical issuance of the UASL under NTP 1999 regime in 2008. On 14 December 2005, UASL guidelines were issued that provided for grant of licences without any restriction or cap in a service area on FCFS basis. 159 licences were issued and there were 53 UASL applications pending when TRAI made reference to DoT. On 17 July 2007, the MoC&IT gave his approval for the proposal of the department.[82]

While rendering his opinion in the context of Migration Package, the AG had proposed that the effective date of licensees of Basic and Cellular Operators might be extended across the board in all cases, by a period ranging from a minimum of six months to a maximum of nine months as

---

[80]Ibid., pp. 52–56.
[81]Ibid., p. 62.
[82]Ibid., p. 67

the government may decide. This proposal was based on the reason that notwithstanding the fact that the licensees could not as of legal right claim extension of effective dates, but in view of the time taken in various governmental clearances like Standing Advisory Committee on Frequency Allocations, Foreign Investment Promotion Board, etc., some relief was called for. In the note for the Cabinet dated 28 June 1999, the DoT submitted that this extension in effective date will involve a revenue loss of approximately ₹1,443.58 crore to the government by way of reduction in the receipt of licence fee if the extension were to be granted for six months and ₹2,156.70 crore, if it were for nine months. It was mentioned that this action could invite criticism from audit as well as from the public. However, in keeping with the advice of the AG, as also the recommendations of the financial institutions, and the fact that in some cases, extension of effective date had been agreed by the Government, it was proposed in the Cabinet note that extension of effective date by a period of six months might be considered across the board for all licences for cellular services in Telecommunication Circles (with the exclusion of metros) as also in the case of BTS. The proposal was approved by the Cabinet.

While justifying the aforesaid proposal, the then Secretary informed the Committee: 'There was a cost to it. And that cost the Government considered consciously. Yes, it was a cost worthwhile to be met. There were options. We chose an option which may have some cost, we supported that option. But it gave us rich dividends.'[83]

On 28 August 2007, the TRAI gave its recommendations

---

[83]Ibid., p. 32.

on review of licence terms and conditions and capping of number of Access Service Providers. Salient recommendations of the TRAI were as follows:

(i) No cap be placed on the number of Access Service Providers in any Service Area.
(ii) The DoT should examine the issue early and specify appropriate licence fee for UAS licensees who do not wish to utilize the spectrum.
(iii) In future, all spectrum, excluding the spectrum in 800, 900 and 1,800 bands should be auctioned so as to ensure efficient utilization of this scarce resource.
(iv) A licensee using one technology may be permitted up on request, usage of alternate technology and thus allocation of dual spectrum. However, such a licensee must pay the same amount of fee which has been paid by existing licensees using the alternate technology or which would be paid by a new licensee going to use that technology.
(v) Revised spectrum allocation criteria.

On 21 September 2007, an internal committee was constituted to examine the recommendations of TRAI.

A press release was issued notifying the cut-off date for applications for UASLs on 24 September 2007. The internal committee of DoT gave its approval to examine TRAI's recommendations on 28 August 2007. While considering TRAI's recommendations, the Telecom Commission made some changes on 17 October 2007. The press note dated 19 October 2007 was also considered for allocation from the date of payment of prescribed fee.

Four companies were providing CDMA-based mobile service under UASL. On 17 October 2007, DoT took the decision for use of alternate technology based on the TRAI recommendations.

The DoT had received 575 applications for UASL. On 2 November 2007, based on the noting recorded, MoC&IT stated 'LoI may be issued to the applicants received up to 25 September 2007.'[84] On 7 November 2007, the Minister gave his approval to DoT for issue of new UASL. The prime minister acknowledged the letter of the minister on 21 November 2007.

The DoT proceeded to implement the recommendation of TRAI on subscriber linked Spectrum Allocation criterion for CMTS/UAS licensees. It allotted initial spectrum under dual technology policy. It had been implementing a policy of FCFS for grant of UASL. The challenge to the criteria for additional GSM Spectrum, crossover technology, spectrum allocation, the issue of new telecom licences, etc., were addressed in the note of the Solicitor General of India.

On 10 January 2008, the DoT issued a press note on its website notifying that all the applicants were eligible to be issued LoI on FCFS, up to 25 September 2007. The committee desired to know whether the former SG had agreed to this amendment in the press release.

On 10 January 2008, the DoT issued a second press release at their website at 2.45 p.m. asking UASL applicants to depute their authorized representatives to collect the DoT

---

[84]Ibid., pp. 68–71.

responses on the same day.[85]

The Committee noted that even though the decision to advance the cut-off date for receiving UASL applications was taken by DoT on 2 November 2007, a press release to this effect was issued only on 10 January 2008. In this context, in evidence, the Committee specifically desired to know from Siddharth Behura, the former Secretary, as to why this decision was not put out in the public domain at the earliest, as was done on 25 September 2007 following a decision taken on 24 September 2007 about fixing a cut-off date.

In response, the former Secretary submitted that it could not have been notified earlier:[86]

> Since a decision had been taken on 2 November 2007 to have it on 25 September 2007 and reiterated on 4 December 2007, they should have made the notification then if they have not done it. If one does it as early as possible, it should not be done later as you should do it as soon as the decision is taken if there is any delay.
>
> Above the date of issue, four special counters were set up for the issue of LoIs to various groups of companies. Issuance of 120 UASLs took place between February and March 2008. Two more LoIs were issued in July 2008 and these UASLs were signed in August 2008.
>
> Spectrum was allotted to these licensees from April 2008 to May 2009.
>
> Developments Post-Issuance of UASLs were

---

[85]For full text of the press release, see p. 62.
[86]Ibid., p. 95.

submitted in a note dealing with Spectrum issues to the Prime Minister.

The Seventh Chapter explains the assessment of the financial impact on account of the issue of licences and spectrum as assessed by the CAG report. Underpricing of 2G at an entry fee of ₹1,658.5 crore and consequent loss, value based on prices discovered for 3G Spectrum, the sale of equity by some UAS licensee firms at higher value, and the offer of S Tel to pay higher revenue were the three indicators the audit had assessed to determine what they called presumptive loss. The CAG report had clarified that the offer made by S Tel to pay additional revenue share of ₹6,000 (enhanced later to ₹13,752 crore) was only an indicator of the market presumption and it was an attempt to put the price discovery in spotlight through market mechanism with higher value and increased receipts for government.

The Committee examined the concept of presumptive loss and concluded that the very concept of calculation of presumptive loss in the context of allocation of licences and spectrum is misleading.

The TRAI's views regarding the loss and its response to the CBI observed that the authority had issued its recommendations on Spectrum Management and Licensing Framework in May 2010. It was also of the view that telecom services and spectrum should not be treated as a source of revenue for the government. Article 39(b) of the Constitution under Directive Principles of State Policy, inter alia, provides that the 'state shall, in particular, direct its policy towards securing that the ownership and control of the material resources of the community are so distributed as best to sub serve the common good.'

It is helpful to look back to 2002, by when telephones were made available on demand with the help of NTP 1999. This was a far cry from the time that customers had to wait indefinitely to get a connection and even seek connections through the MP quota. In the Tenth Plan document, the Planning Commission defined the meaning of opportunity cost as the cost of any activity measured in terms of value and not chosen. Promotional in nature, pricing and allocation, optimal utilization, opportunity cost, funds for civilian purposes and spectrum pricing are some of the guiding principles of spectrum policy. Even the Advisory Opinion of the Supreme Court on Presidential Reference regarding allocation of natural resources holds that in all circumstances and for all sectors it is neither possible nor optimal to auction natural resources. Yet the accused licensees had to face a prolonged trial to prove their innocence in the eyes of the law.

The Eighth Chapter discusses the role of TRAI in considerable detail. According to this chapter, the telecommunications sector was a state monopoly till the year 1991. However, the new economic policy of 1991 liberalized, privatized and globalized the economy. In 1992, tenders were issued for grant of two licences in four metros. Private operators were granted licences in November 1994.

The need for an independent regulator was felt with the entry of a mixed environment including multiple service providers in order to prevent profiteering; the provision of universal service was in order to protect customers, employees and the environment from the damage of inappropriate behaviour of firms.

An independent telecom regulatory authority was proposed as non-statutory body by the government. On 25 January 1997, an ordinance was passed establishing TRAI as an independent regulatory agency for the telecom sector. TRAI Act 1997 was enacted by Parliament on 28 March 1997 and amended on 25 March 2000. The TDSAT came into existence on 29 May 2000.

The Ninth Chapter explains the meaning of 'spectrum' and demonstrates its role and importance as a precious national asset. It explicates that radio frequency spectrum refers to the collection of various types of electromagnetic radiations of different wavelengths. Spectrum is a radio frequency on which all communication signals travel. Radio frequency spectrum is the entire range of wavelengths of electromagnetic radiation which is used as carrier of wireless transmission and thus, a basic requirement for providing wireless services. It is a finite but non-consumable global natural resource and commands high economic value in the telecommunication sector. The spectrum is used and not consumed, and it is wasted if not optimally and efficiently used. It cannot be owned but used and shared amongst various countries, services, users, technologies, etc., without any element of exclusiveness. Radio communication networks are like global society necessitating appropriate discipline.[87]

Utilization (assignment of frequency) of these resources follows laws of physics and is governed by international

---

[87]Ibid., pp. 123–146.

treaties notably, the Constitution, the Convention and Radio Regulations of the ITU. India falls in the ITU region.

Spectrum is a precious national asset which is shared by various government and private user organizations like defence, police, intelligence and other security agencies, public telecommunications, broadcasting, railways, public utility organizations, oil and electricity grids, atomic energy, mining and steel, shipping and airlines, as well as private and public telecom operators, for a variety of applications mainly, public telecom services, aeronautical and maritime safety communications, radars, seismic surveys, rocket and satellite launching, earth exploration and forecasting of natural calamities.

As mobile services need spectrum, sufficient spectrum must be made available for successful implementation of wireless based telecom infrastructure. In this connection, the Government formulated the NFAP, which is based on International Radio Regulations for optimal use of the scarce resource. The NFAP forms the basis for development and manufacturing of wireless equipment and spectrum utilization in the country.

The NFAP was established in 1981 and has been modified from time to time, its latest revision being in 2008. The NFAP 2008 has become effective from 1 April 2009 which is presently in force.

In India, mobile services which use GSM technology work in the frequency bands of 900 and 1,800 MHz and those in CDMA technology work in the 800 MHz band. Earlier, the 800, 900 and 1,800 MHz bands were allotted to defence services for their mobile communication usage.

Presently, 25 MHz spectrum in 900 MHz band and 75 MHz in the 1,800 MHz band is earmarked for GSM services. For CDMA services, 20 MHz spectrum in the 800 MHz band is earmarked.

The JPC Report thus provided a broad overview of the mobile telephony sector through the tenures of several governments. Although faults were found at different stages of the evolving policies, nowhere did the Report suggest culpability that was of criminal nature. At best it could be said that the government of the day responded to the experience of the industry keeping the ultimate policy goals in mind. It was somewhat of a learning experience in the unique Indian market and context distinguishable from the experience of most other countries. Although the Report consciously underscored that the then prime minister fulfilled his responsibility to caution and advise the Minister and it would hardly be fair to have expected more from him, the latter was not found to have deliberately misled the PM as some had suggested. There, certainly, are useful lessons in the findings on the difficulties of Cabinet form of government with the doctrine of collective responsibility.

6
DARK CLOUDS OVER SUNSHINE SECTOR

If any statement continued to ring in one's ears months after the hurly burly of the 2G implosion, it is the phrase used by my colleague and the then replacement Telecom Minister Kapil Sibal—'zero loss'. The assertion was made on very sound reasoning based on the Planning Commission Working Paper for the Tenth Five Year Plan (2002–2007) that identified territory penetration rather than revenue earning as national priority. But sadly, the statement turned out to be a red rag to the bull. A howl of protest went up in Parliament and outside, driving Sibal for cover. What was indeed the truth spoken with responsibility became impudent denial for the public inquisition and we left the space undefended in the hope of cutting our losses, individually and collectively. The judgment of Judge O.P. Saini came as great relief to all of us and of course to Kapil Sibal who lost a bit of time reiterating his red rag to the BJP bull. But, in a sense, this is a case of history vindicating us rather than being a valiant fight to overcome adversity.

Sadly, given the system under which we worked, there was seldom any occasion for the entire Cabinet to discuss the implications of the 2G affair. It remained the domain of the telecom minister, first the elusive and besieged A. Raja, in the eye of the storm with few friends beyond the DMK, and then Sibal, the chosen gladiator, fighting a Teflon battle to clean up the mess. Our recognized troubleshooter, P. Chidambaram appeared to stay at arm's length because somewhat unfair attempts were being made to somehow embroil him in the controversy. His involvement had only been to seek clarification on pricing and, upon Raja's insistence, flagging future allocation pricing. A completely uncalled for controversy was raised about the explicit approval of the 2001 pricing given by Chidambaram and the prime minister. The matter caused some discomfort amongst Chidambaram and his successor in the Finance Ministry as well as in the PMO. Understandably, everyone was tense about what the courts might do. I visited Chidambaram at his official residence several times in the evenings to reassure him and clear any misgivings. On one such occasion I received a call from the prime minister who was on a foreign tour. That was a fortuitous opportunity to get him to speak to Chidambaram. I stepped out of the room while they were speaking and returned to find Chidambaram at considerable ease. Luckily, both the Supreme Court and the trial court gave short shrifts to attempts of dragging Chidambaram's name. I am glad that I could do a bit to repair the situation. But the very day we ourselves concluded that the problem was not a collective one but that of an individual minister, we lost the battle of perception. In the end, ours was a defeat of perception rather than of actual failure.

The only time there was some discussion in the Cabinet on 2G was after Justice G.S. Singhvi's Bench of the Supreme Court had handed down its judgement cancelling all the licences. I was then the minister of law and had worked steadily to smoothen ruffled judicial feathers, particularly to address what appeared to be perceived as trust deficit between Justice Singhvi and Attorney General Ghulam Vahanvati. In one of my meetings with Justice Singhvi, he spoke at length about the unwholesome atmosphere in Delhi unlike Chandigarh where he said judges mixed easily with senior lawyers. He told me about his high regard for Vahanvati as well as for the prime minister and deeply regretted the helplessness with which he felt he had to watch media misreporting since judges do not have a way of responding to it. Justice Singhvi was a strict, moralistic person but extremely gracious and cordial. He flagged some incidents that caused a mistrust between some judges and the government. From the treatment I received from him every time I called him, the judge assured me that some of that mistrust was reduced. Some years later, I had the opportunity to appear before Justice Singhvi as Chairperson of the Competition Appellate Tribunal and found him most affable and caring. When the festival of Id came during a long appeal hearing, he generously invited the entire Bar to join in partaking in sevai I had offered to bring. If this spiritual and generous man thought ill of us at any stage, it is indeed a great misfortune.

The Verdict That Caused an Upheaval

Two writ petitions filed by Dr Subramanian Swamy in 2011 and the Centre for Public Interest Litigation (CPIL) in 2010 represented by Prashant Bhushan challenged the 2008 allotment of 2G licences. The questions which arose for consideration in the writ petitions were:[88]

1. Whether the Government has the right to alienate, transfer or distribute natural resources/national assets otherwise than by following a fair and transparent method consistent with the fundamentals of the equality clause enshrined in the Constitution.
2. Whether the recommendations made by the TRAI on 28 August 2007 for grant of UASL with 2G spectrum in 800, 900 and 1,800 MHz at the price fixed in 2001, which were approved by the DoT, were contrary to the decision taken by the Council of Ministers on 31 October 2003.
3. Whether the exercise undertaken by DoT from September 2007 to March 2008 for the grant of UASL to the private respondents in terms of the recommendations made by TRAI is vitiated due to arbitrariness and malafides and is contrary to public interest.
4. Whether the policy of FCFS followed by DoT for grant of licences is ultra vires the provisions of Article 14 of the Constitution and whether the said principle was arbitrarily changed by the MoC&IT, without consulting the TRAI, with a view to favour some of the applicants.

---

[88]Centre for Public Interest Litigation and others vs. Union of India and others; Writ Petition (Civil) No. 423 of 2010, http://www.supremecourtofindia.nic.in/jonew/judis/39041.pdf

5. Whether the licences granted to ineligible applicants and those who failed to fulfil the terms and conditions of the licence are liable to be quashed.

On 2 February 2012, the Supreme Court ruled on cancelling all 122 spectrum licences granted during Raja's term as MoC&IT minister. The judgement described the allocation of 2G spectrum as 'unconstitutional and arbitrary'. The Bench of Justices—G.S. Singhvi and Asok Kumar Ganguly—imposed a fine of ₹50 million ($7,80,000) on Unitech Wireless, Swan Telecom and Tata Teleservices and a ₹5 million ($78,000) fine on Loop Telecom, S Tel, Allianz Infratech and Sistema Shyam Tele Services. According to the ruling, the current licences would remain in place for four months, after which the government would reissue the licences.

In the ruling, the Supreme Court Bench maintained that the 'exercise undertaken by the officers of DoT between September 2007 and March 2008, under the leadership of the then MoC&IT was wholly arbitrary, capricious and contrary to public interest apart from being violative of the doctrine of equality. The material produced before the court shows that the minister wanted to favour some companies at the cost of the public exchequer and for this purpose, he took the following steps:

 i. Soon after his appointment as MoC&IT, he directed that all the applications received for grant of UASL should be kept pending till the receipt of the TRAI recommendations.
 ii. The recommendations made by TRAI on 28 August 2007 were not placed before the full Telecom

Commission which, among others, would have included the Finance Secretary. The notice of the meeting of the Telecom Commission was not given to any of the non-permanent members despite the fact that the recommendations made by TRAI for allocation of spectrum in 2G bands had serious financial implications. This has been established from the pleadings and the records produced before this Court which show that after issuance of licences, three applicants transferred their equities for a total sum of ₹24,493 crore in favour of foreign companies. Therefore, it was absolutely necessary for the DoT to take the opinion of the Finance Ministry, as per the requirement of the Government of India (Transaction of Business) Rules, 1961.

iii. The officers of DoT who attended the meeting of the Telecom Commission held on 10 October 2007 hardly had any choice but to approve the recommendations made by the TRAI. If they had not done so, they would have incurred the wrath of the minister.

iv. In view of the approval by the Council of Ministers of the recommendations made by the GoM in 2003, the DoT had to discuss the issue of spectrum pricing with the Ministry of Finance. Therefore, the DoT was under an obligation to involve the Ministry of Finance before any decision could be taken in the context of paras 2.78 and 2.79 of the TRAI's recommendations. However, as the minister was very much conscious of the fact that the secretary, Finance, had objected to the allocation of 2G Spectrum at the rates fixed in 2001, he did not consult the finance minister or the officers of the Ministry of Finance.

v. The MoC&IT brushed aside the suggestion made by the Minister of Law and Justice for placing the matter before the EGoM. Not only this, within a few hours of the receipt of the suggestion made by the prime minister in his letter dated 2 November 2007 that keeping in view the inadequacy of spectrum, transparency and fairness should be maintained in the matter of allocation thereof, the minister rejected the same by saying that it will be unfair, discriminatory, arbitrary and capricious to auction the spectrum to new applicants because it will not give them a level playing field.

vi. The minister introduced the cut-off date as 25 September 2007 for consideration of the applications received for grant of licence despite the fact that only one day prior to this, a press release was issued by DoT fixing 1 October 2007 as the last date for receipt of the applications. This arbitrary action of the minister, although innocuous in appearance actually benefited some of the real estate companies who did not have any experience in dealing with telecom services and who had made applications only on 24 September 2007, i.e. one day before the cut-off date fixed by the minister on his own.

vii. The cut-off date, i.e. 25 September 2007 decided by the minister on 2 November 2007 was not made public till 10 January 2008 and the FCFS policy which was being followed since 2003 was changed by him on 7 January 2008 and was incorporated in the press release dated 10 January 2008. This enabled some of the applicants, who had access either to the minister or the officers of the DoT to get the demand drafts, bank guarantee, etc., prepared in advance for compliance with conditions of the LoIs, which was the basis for determination of seniority for grant of licences and allocation of spectrum.

viii. The meeting of the Full Telecom Commission, which was scheduled to be held on 9 January 2008 to consider issues relating to the grant of licences and pricing of spectrum was deliberately postponed on 7 January 2008 so that the secretary, Finance, and secretaries of three other important departments would not be able to raise objections against the procedure devised by the DoT for grant of licence and allocation of spectrum by applying the principle of a level playing field.

ix. The manner in which the exercise for grant of the LoIs to the applicants was conducted on 10 January 2008 leaves no room for doubt that everything was stage-managed to favour those who were able to know in advance the change in the implementation of the FCFS policy. As a result of this, some of the companies which had submitted applications in 2004 or 2006 were pushed down in the list of priority and those who had applied between August and September 2007 succeeded in being entitling for allocation of spectrum on high priority.

Curiously, the complaint was made by Subramanian Swamy rather than a disappointed applicant. The Tamil Nadu politics angle was more than apparent in the exercise.

The Court answered the questions raised in the writ petitions as follows:

Re Question 1:
The State is the legal owner of the natural resources as a trustee of the people and although it is empowered to distribute the same, the process of distribution must be guided by the constitutional principles including the doctrine of equality and larger public good.

Re Question 2:
The recommendations made by the TRAI were flawed in many respects and implementation thereof by the DoT resulted in gross violation of the objective of NTP 1999 and the decision taken by the Council of Ministers on 31 October 2003.

Although while making recommendations on 28 August 2007, the TRAI had itself recognized that spectrum was a scarce commodity, it made recommendation for allocation of 2G Spectrum on the basis of 2001 price by invoking the theory of a level playing field. Para 2.40 of the recommendations dated 28 August 2007 shows that as per TRAI's own assessment, the existing system of spectrum allocation criteria, pricing methodology and the management system suffered from a number of deficiencies and there was an urgent need to address the issues linked with spectrum efficiency and its management and yet it decided to recommend the allocation of spectrum at the price determined in 2001. All this was done in the name of growth, affordability, penetration of wireless services in semi-urban and rural areas, etc. Unfortunately, while doing so, the TRAI completely overlooked that one of the main objectives of the NTP 1999 was that spectrum should be utilized efficiently, economically, rationally and optimally and there should be a transparent process of allocation of frequency spectrum. In terms of the decision taken by the Council of Ministers in 2003 to approve the recommendations of the GoM, the DoT and Ministry of Finance were required to discuss and finalize the spectrum pricing formula.

To say the least, the entire approach adopted by the TRAI was lopsided and contrary to the decision taken by the Council of Ministers; its recommendations became a handle for the then MoC&IT and the officers of DoT who virtually gifted away the important national asset at throwaway prices by wilfully ignoring the concerns raised from various quarters including the prime minister, the Ministry of Finance and also some of its own officers. This becomes clear from the fact that soon after obtaining the licences, some of the beneficiaries offloaded their stakes to others in the name of transfer of equity or infusion of fresh capital by foreign companies, and thereby made huge profits. We have no doubt that if the method of auction had been adopted for grant of licence which could be the only rational transparent method for distribution of national wealth, the nation would have been enriched by many thousand crores.

While it cannot be denied that TRAI is an expert body assigned with important functions under the 1997

Act, it cannot make recommendations overlooking the basic constitutional postulates and established principles and thereby deny the people from participating in the distribution of national wealth and benefit a handful of persons. Therefore, even though the scope of judicial review in such matters is extremely limited, as pointed out in Delhi Science Forum vs. Union of India [(1996) 2 SCC 405] and a large number of other judgements relied upon by the learned counsel for the respondents, keeping in view the facts which have been brought to the notice of the Court that the mechanism evolved by TRAI for allocation of spectrum and the methodology adopted by the then MoC&IT and the officers of the DoT for grant of UASL may have caused huge loss to the nation, we have no hesitation to record a finding that the recommendations made by TRAI were flawed in many respects, their implementation by the DoT resulted in gross violation of the objective of the NTP 1999 and the decision taken by the Council of Ministers on 31 October 2003.

We may also mention that even though in its recommendations dated 28 August 2007, the TRAI had not specifically recommended that entry fee be fixed at 2001 rates, Para 2.73 and other related paragraphs of its recommendations state that it has decided not to recommend the standard option for pricing of spectrum in 2G bands keeping in view a level playing field for the new entrants. It is impossible to approve the decision taken by the DoT to act upon those recommendations. We also consider it necessary to observe that in today's dynamism and unprecedented growth of telecom sector, the entry fee determined in 2001 ought to have been treated by TRAI as wholly unrealistic for grant of licence along with start-up spectrum. In our view, the recommendations made by TRAI in this regard were contrary to the decision of the Council of Ministers that DoT shall discuss the issue of spectrum pricing with the Ministry of Finance along with the issue of incentive for efficient use of spectrum as well as disincentive for suboptimal usages. Being an expert body, it was incumbent upon the TRAI to make suitable recommendations even for the 2G bands especially in light of the deficiencies of the present system which it had itself pointed out. We do not find merit in the reasoning of the TRAI that the consideration of maintaining a level playing field prevented a realistic reassessment of the entry fee.

Re Questions 3 and 4:
There is a fundamental flaw in the FCFS policy as it involves an element of pure chance or accident. In matters involving award of contracts or grant of licence or permission to use public property, the invocation of the FCFS policy has inherently dangerous implications. When it comes to alienation of scarce natural resources like spectrum, etc., it is the burden of the State to ensure that a non-discriminatory method is adopted for distribution and alienation, which would necessarily result in protection of national/public interest. A duly publicized auction conducted fairly and impartially is perhaps the best method for discharging this burden and the methods like FCFS when used for alienation of natural resources/public property are likely to be misused by unscrupulous people who are only interested in garnering

maximum financial benefit and have no respect for the constitutional ethos and values.

Re Question 5:
122 licences were quashed and additional penalties were imposed on those service providers whose conduct was found questionable by the court.

When it is clearly demonstrated that the policy framed by the State or its agency/instrumentality and/or its implementation is contrary to public interest or is violative of the constitutional principles, it is the duty of the court to exercise its jurisdiction in larger public interest and reject the stock plea of the State that the scope of judicial review should not be exceeded beyond the recognized parameters.

The writ petitions were thus allowed as follows:[89]

   i   The licences granted to the private respondents on or after 10 January 2008 pursuant to two press releases issued on 10 January 2008 and subsequent allocation of spectrum to the licensees are declared illegal and are quashed.

   ii   The above direction shall become operative after four months.

   iii   Keeping in view the decision taken by the Central Government in 2011, TRAI shall make fresh recommendations for grant of licence and allocation of spectrum in 2G band in 22 Service Areas by auction, as was done for allocation of spectrum in 3G band.

---

[89] https://indiankanoon.org/doc/70191862/

iv The Central Government shall consider the recommendations of TRAI and take appropriate decision within next one month and fresh licences be granted by auction.

v Respondent Nos. 2, 3 and 9 who have been benefited at the cost of Public Exchequer by a wholly arbitrary and unconstitutional action taken by DoT for grant of UASL and allocation of spectrum in 2G band and who offloaded their stakes for many thousand crores in the name of fresh infusion of equity or transfer of equity shall pay a cost of ₹5 crores each.[90] Respondent Nos. 4, 6, 7 and 10 shall pay a cost of ₹50 lakhs each because they too had benefited by the wholly arbitrary and unconstitutional exercise undertaken by DoT for the grant of UASL and allocation of spectrum in 2G band.[91] We have not imposed cost on the respondents who had submitted their applications in 2004 and 2006 and whose applications were kept pending till 2007.

vi Within four months, 50 per cent of the cost shall be deposited with the Supreme Court Legal Services Committee for being used for providing legal aid to poor and indigent litigants. The remaining 50 per cent cost shall be deposited in the funds created for Resettlement and Welfare Schemes of the Ministry of Defence.

vii However, it is made clear that the observations made

---

[90]https://indiankanoon.org/doc/70191862/
[91]Ibid.

in this judgment shall not, in any manner, affect the pending investigation by the CBI, Directorate of Enforcement and other agencies or cause prejudice to those who are facing prosecution in the cases registered by the CBI or who may face prosecution on the basis of chargesheet(s) which may be filed by the CBI in future and the Special Judge, CBI shall decide the matter uninfluenced by this judgment. We also make it clear that this judgment shall not prejudice any person in the action which may be taken by other investigating agencies under Income Tax Act, 1961, Prevention of Money Laundering Act, 2002 and other similar statutes.

Although the policy for awarding licences was FCFS, Raja was alleged to have changed the rules so it applied to compliance with conditions instead of the application itself. Furthermore, on 10 January 2008, companies were given only a few hours to collect LoI and make payments; some executives were allegedly tipped off and arrived ahead of their competitors with banker's drafts.

Blow to the Mobile Telephony Industry

The 2G judgement was pronounced on the day when the second judge, Justice A.K. Ganguly was to retire. The city was rife with rumours and innuendoes for weeks. The night before the judgement day someone left a copy at my gate in a sealed envelope. I was shocked, to say the least, but there was little I could do in the middle of the night. Early next morning, I was scheduled for a trip to a southern state. I went through the business of the day, but asked for updates from Delhi. Some parts of the judgement as delivered in the court seemed different from the copy I had received. Upon my return to Delhi, I discovered that the judgement uploaded was different. I immediately brought the matter up with Chief Justice Kapadia who promised to have the entire matter examined. Although I did not hear anything further, I am given to understand that some action was taken against a clerk of the court registry.

People might well have forgotten that there was a time when Justice G.S. Singhvi's Bench was not only hearing the main 2G matter but was also monitoring the investigation. The Delhi High Court had refused to go the entire distance in the PIL filed before it, causing the petitioners to rush to the Supreme Court. All courts, including the Delhi High Court whose constitutional duty was to supervise trial courts of Delhi, were directed not to hear any matter that was related to 2G. All bail applications therefore had to come directly to the Supreme Court and Section 482 became a dead letter. The Supreme Court even appointed the Special Public Prosecutor and upon the incumbent's elevation to the Bench, chose his successor, Senior Advocate Anand Grover.

After Chief Justice Kapadia retired, unsuccessful attempts were made by parties to challenge the orders of Justice Singhvi's court relating to the trial proceedings in the hope that the new Chief Justice Altamas Kabir might intervene. But that petition too was ultimately placed before Justice Singhvi and never reached a decision.

On 3 August 2012, after the Supreme Court directive, the

government revised the base price for 5MHz 2G spectrum auctions to ₹140 billion ($2.2 billion), raising its value to about ₹28 billion ($440 million) per MHz (near the CAG's estimate of ₹33.5 billion [$520 million] per MHz). It is another matter that to a large extent there were no takers at that price and the reserve price had to be revised downwards. Owing to the uncertainty caused by the Supreme Court decision, the disqualification of certain key players and paucity of bank funding available, the government was forced to auction the spectrum in instalments in order to balance supply and demand. The outcome of the first several rounds of auction was a clear indication of the enormous stress suffered by the sunshine sector.

The cancellation of the licences virtually pulled the rug from under the feet of the mobile telephony industry. Understandably, the court gave some time to unravel the existing system and conduct fresh auctions. The uncertainty about the future and the reluctance of banks to give further support hit productivity. In view of the fact that spectrum might need to be allocated to individual entities from time to time in accordance with the criteria laid down by the government, such as subscriber base, availability of spectrum in a particular circle, inter se priority depending on whether the spectrum comprises the initial allocation or additional allocation, etc., it was felt that it may not always be possible to conduct an auction for the allocation of spectrum.

In view of the ground situation, the auctioning of spectrum in the 2G bands was to result in a situation where none of the licensees using the 2G bands of 800 MHz, 900 MHz and 1,800 MHz would have paid any separate upfront fee for the allocation of spectrum. Meanwhile, the government received various notices from companies based in other countries, invoking bilateral investment agreements and seeking damages against the Union of India by reason of the cancellation/threat of cancellation of the licences. Many of those have matured into international arbitrations in which we have little chance of averting huge damages. The Ministry of Finance, once again under Chidambaram, reacted by deciding to renegotiate all Bilateral Investment Treaties (BITs), the immediate fallout of which was the treaty with Mauritius. But implications of the 2G judgement included actions by affected parties under the extant terms of BITs, still pending before International Arbitrators, which include Khaitan Holdings/India–Mauritius BIPA, India–UK BIPA and Sistema Joint Stock Financial Corporation/India–Russia BIPA.

Related Dimensions

Malaysia's Astro Asia Networks Ltd. (AAANL), which has been charge sheeted by the CBI in the Aircel-Maxis case, said it has neither been formally intimated nor served with any charge sheet; they called the CBI allegations baseless. Astro All Asia and other companies held shares in Sun Direct TV Private Limited. It appears that the matter arose out of the complaint of former Aircel Chief Mr Sivasankaran (Siva) in relation to the sale of his company, Aircel Limited (Aircel) to Maxis Communications Berhad, a Malaysian Company. This complaint led to CBI investigations into the alleged misdemeanours of Dayanidhi Maran, which had allegedly implicated Astro Holdings, Astro All Asia and SAHEL companies in a false case. The dispute notice has been sent to Ministry of Information and Broadcasting for further processing.

Not having responded to the notice of arbitration on time, the claimant had approached the president of the International Court of Justice (ICJ) for appointment of an arbitrator on behalf of the government. Later, the government appointed a counsel team who, in consultation with the GoI, nominated its arbitrator. Preliminary request for interim measures by the claimants was denied. However, they have again approached the tribunal for interim measures due to the recent judgement of the CBI court discharging the Marans. Further, this matter has been combined with the SAHEL arbitration and hearings are scheduled for 2018.

The dispute arose from the judgement and pursuant auction by the DoT for allocation of spectrum in 2013 and 2015 and the related guidelines issued by the DoT which were challenged by Sistema as being discriminatory and arbitrary. The TDSAT had ruled in favour of Sistema and efforts to settle the dispute were made.[92] GoI (DoT) has filed an appeal which is pending in the Supreme Court. In addition to Batelco's exit on 21 February 2012, Telenor (the majority shareholder in Uninor) terminated its agreement with Unitech and sued it for 'indemnity and compensation'. On 23 February 2012, Etisalat DB Telecom sued DB Realty corporate promoters Shahid Balwa and Vinod Goenka for fraud and misrepresentation.

---

[92] www.financialexpress.com/india-news/sc-stays-tdsat-order-in-sistema-shyam-case/841355/

An add-on to the main 2G show was the controversy involving the Aircel-Maxis deal. On 6 June 2011, Sivasankaran complained to the CBI about not receiving a telecom licence and being forced by Telecom Minister Dayanidhi Maran to sell Aircel to Malaysia-based Maxis Communications group.[93] The licences were allegedly issued after the deal was made. Sivasankaran also alleged that brothers Dayanidhi and Kalanithi Maran received kickbacks in the form of investments by the Maxis group through the Astro network in Sun TV Network owned by the Maran family. In the wake of the allegations, Maran resigned on 7 July.

On 10 October, CBI registered a case and raided properties owned by the Marans. CBI sources said that although no evidence of coercion was found in the Aircel sale, they found substantial evidence that Maran had favoured the company's takeover by Maxis and deliberately delayed Sivasankar's files. On 8 February 2012, the Enforcement Directorate (ED) registered a money-laundering case against the Maran brothers for allegedly receiving illegal compensation of about ₹5.5 billion in the Aircel-Maxis deal.[94]

During the CBI probe, Sivasankaran maintained that the Maran brothers had forced him to sell his 74 per cent share in Aircel to Maxis by threatening his life, giving the CBI a list of over 10 witnesses. In September 2012, the CBI

---

[93]https://www.livemint.com/Home-Page/Yz33IdNBnNTCAdxzkQormK/2G-Scam--Sivasankaran-to-CBI-Maran-forced-me-to-sell-Airce.html

[94]https://economictimes.indiatimes.com/news/politics-and-nation/dayanidhi-maran-all-accused-discharged-in-aircel-maxis-case/articleshow/56935208.cms

said it had completed its investigations in India and was awaiting the response to a letter rogatory sent to Malaysia and a questionnaire from Malaysian businessman T. Ananda Krishnan before filing a charge sheet. On 29 August 2014, the CBI filed a charge sheet against the Maran brothers, Krishnan, Malaysian national Augustus Ralph Marshall and six others apart from four firms—Sun Direct TV Pvt. Ltd., Maxis Communication Berhad, Astro All Asia Network PLC and South Asia Entertainment Holding Ltd. as accused in the case. On 29 October 2014, special CBI judge O.P. Saini said that he found enough evidence to proceed with the prosecution and hence summoned former Telecom Minister Dayanidhi Maran and others as accused. Based on the CBI chargesheet, the ED, on 1 April 2015, attached properties of the Maran brothers worth ₹742 crore. Yet the court of Judge Saini finally discharged the Maran brothers and the two companies M/s Sun Direct TV (P) Ltd. (SDTPL) and M/s South Asia Entertainment Holdings Ltd.

Subramanian Swamy further alleged that in 2006, a company controlled by Karti Chidambaram, son of former Finance Minister P. Chidambaram, received a 5 per cent share of Aircel to get part of ₹40 billion paid by Maxis Communications for the 74 per cent stake of Aircel. According to Swamy, Chidambaram withheld Foreign Investment Promotion Board (FIPB) clearance of the deal until his son received the 5 per cent share in Siva's company. The issue was raised a number of times in Parliament by the Opposition, which demanded Chidambaram's resignation. Although he and the government denied the allegations, *The Pioneer* and *India Today* reported the existence of documents showing that Chidambaram delayed approval of the FDI proposal by about seven months. https://www.indiatoday.in/india/north/story/document-shows-chidambaram-delayed-aircel-maxis-deal-101532-2012-05-08

Chidambaram has answered the allegations several times and forcefully even going to the extent of telling Parliament that he preferred someone thrusting a dagger in his back than to question his integrity.[95] The decisions were taken by the FIPB, and the finance minister merely counter signs it. Yet Karti remains in the eye of the storm as the agencies continue to doggedly pursue him. It is more than obvious that the incumbent government's intention is to embarrass the Congress party's leading administrative and economics expert. Fortunately, he remains resolute and determined to expose the development record of the BJP.

Chidambaram has said several times that the deal was sent for approval to him by the FIPB, which reports to the Ministry of Finance. FIPB consists of five secretaries to the Government of India who submitted the case to the finance minister and sought approval. As finance minister, he granted approval in the normal course of business.

The Price Still Remains to be Paid

The 2G saga has thrown up interesting dimensions of Indian polity. To begin with, after all this time, it is far from clear what the judiciary's ultimate role will be. The Supreme Court, after entertaining the PIL and directing a court-monitored

---

[95] https://www.firstpost.com/business/chidu-karti-maxis-link-the-dotted-line-is-very-faint-306459.html

investigation, went to the extent of appointing a Special Public Prosecutor in the normal course duty of the State government. It even went on to persuade senior counsel V. Venugopal not to recuse on the ground of having been engaged for a private party in the matter at some stage. Thereafter, the court virtually approved the charge sheet and even prohibited the High Court from entertaining any proceedings relating to the trial. Even bail matters were directed to be heard only by the Supreme Court. Several trial-related matters remained pending for months on end because the court's calendar did not permit it to take them up.

Although the matters are still sub judice, one thing might well be said—we have completely obliterated the idea of an honest mistake in official decision-making. The entire jurisprudence of judicial review of administrative action is based on illegality of decisions on various grounds such as ultra vires, mala fides, non-application of mind, failure of natural justice et al. Every day, decisions are set aside by courts and fresh consideration is directed. In extreme cases, costs or ad hoc penalties are imposed. But it is far from criminal action. The recent proceedings in 2G and the coal allocation matters are mistakes which are made into criminal liability without further thought. The basis of the criminal proceedings relates ultimately to the auction of public resources. Justice Singhvi proceeded on the basis that not holding an auction was per se unlawful but the five-judge Constitutional Bench held the opposite. What is an honest civil servant to do in such circumstances?

While a moralistic approach is quite obvious, particularly

with the profound principles espoused by the court, the fact remains that the same court has dwelt at length on the alternative economic analysis of law which in this case might have given an entirely different outcome. Some years later, this model of judicial making was used for the first time by Justice A.K. Sikri in Shiva Shakthi Sugar Ltd. vs Shri Renuka Sugar Mills Ltd., though it is yet to be endorsed by other Justices of the court.

Finally, one might legitimately ask how many different scrutinies the 2G episode will go through for a closure. The Supreme Court Bench of Justice Singhvi had taken a view with the express caveat that it should not influence the trial court. The Constitutional Bench has taken an entirely different view. The matter had itself surfaced due to the CAG prima facie findings of a flawed decision. The JPC examined the CAG report and came to a divided verdict. In addition, the Justice Shivraj Patil Committee Report submitted its conclusions to the government. We are none the wiser for all that material except to say that some people cleverly built a mountain out of a mole hill. The price was extracted from the Congress party, but an even greater price still remains to be paid by the country.

7
PRESIDENTIAL REFERENCE

The Supreme Court 2G judgement was a bit like a knockout in a boxing bout; we barely made it to the corner on unsteady legs. The implications were far-reaching for the government, the mobile telephony sector, several individuals and the country. There was no point seeking a review of the judgement. After some thought and consultation with Pranab Mukherjee, Kapil Sibal, P. Chidambaram et al, I discussed the matter with the prime minister. The AG was kept in the loop. In the Cabinet meeting where the issue was briefly discussed, I ventured to suggest that other judges of the court might be more sensitive to the government's concerns. Mr Mukherjee reacted by telling me not to indulge in wishful thinking. One could sense the obvious tension between wizened politicians and the judiciary. And yet this was the time that my relations with successive chief justices were excellent. Far from the NDA government's indefinite procrastination of the Supreme Court Collegium's recommendations, I used to get the lists cleared right away. I even had an arrangement with Chief Justice Kapadia that he would give me the names on a slip of paper before the official letter reached the Ministry of Law & Justice.

Questions and Answers with Far Reaching Implications

Under the circumstances, the questions of the far-reaching national and international implications were required to be examined in relation to the conduct of the auction and the regulation of the telecommunications industry in accordance with the Supreme Court Division Bench judgement and FDI in the telecom industry and in other infrastructure sectors. Under Sub-Clause (1) of Article 143 of the Constitution of India, President Pratibha Devisingh Patil referred the following questions to the Constitution Bench of the Supreme Court of India for consideration and report thereon:

Question 1. Whether the only permissible method for disposal of all natural resources across all sectors and in all circumstances is by the conduct of auctions.

Question 2. Whether a broad proposition of law that only the route of auctions can be resorted to for disposal of natural resources does not run contrary to several judgements of the Supreme Court, including those of the larger Benches.

Question 3. Whether the enunciation of a broad principle, even though expressed as a matter of constitutional law, does not really amount to formulation of a policy and has the effect of unsettling policy decisions formulated and approaches taken by various successive governments over the years for valid considerations, including lack of public resources and the need to resort to innovative and different approaches for the development of various sectors of the economy.

Question 4. What is the permissible scope for interference by courts with policymaking by the government including methods for disposal of natural resources?

Question 5. If the court holds within the permissible scope of judicial review, that a policy is flawed, whether the court is not obliged to take into account investments made under the said policy including investments made by foreign investors under multilateral/bilateral agreements.

Question 6. If the answers to the aforesaid questions lead to an affirmation of the judgement dated 2 February 2012, then the following questions may arise, viz.

    i. Whether the Judgement [(2012) 3 SCC 1] is required to be given retrospective effect so as to unsettle all licences issued and 2G Spectrum (800 MHz, 900 MHz and 1,800 MHz bands) allocated in and after 1994 and prior to 10 January 2008.

   ii. Whether the allocation of 2G spectrum in all circumstances and in all specific cases for different policy considerations would nevertheless have to be undone. And specifically, whether the telecom licences granted in 1994 would be affected.

  iii. Whether the telecom licences granted by way of basic licences in 2001 and licences granted between

2003–07 would be affected.
iv. Whether it is open to the Government of India to take any action to alter the terms of any licence to ensure a level playing field among all existing licensees.
v. Whether the Dual Technology licences granted in 2007 and 2008 would be affected.
vi. Whether it is necessary or obligatory for the Government of India to withdraw the spectrum allocated to all existing licensees or to charge for the same with retrospective effect, and if so, on what basis and from what date.

Question 7. Whether, while taking action for conduct of auction in accordance with the orders of the Supreme Court, it would remain permissible for the government to:

a. Make provision for allotment of spectrum from time to time at the auction discovered price and in accordance with the laid down criteria during the period of validity of the auction determined price.
b. Impose a ceiling on the acquisition of spectrum with the aim of avoiding the emergence of dominance in the market by any licensee/applicant duly taking into consideration TRAI recommendations in this regard.
c. Make provision for allocation of spectrum at auction-related prices in accordance with the laid down criteria in bands where there may be inadequate or no competition (for e.g. there is expected to be a low level of competition for CDMA in 800 MHz band and TRAI has recommended an equivalence ratio of 1:5 or 1:3 × 1:5 for 800 MHz and 900 MHz bands

depending upon the quantum of spectrum held by the licensee that can be applied to auction price in 1,800 MHz band in the absence of a specific price for these bands).

Question 8. What is the effect of the judgement [(2012) 3 SCC 1] on 3G Spectrum acquired by entities by auction whose licences have been quashed by the said judgement.

The court was conscious that a bare reading of the reference showed that it was occasioned by the earlier 2G decision of the Division Bench of the Supreme Court. That indeed was the intent of the Cabinet, but by the time it came to be argued before the Constitutional Bench, sensing the judges' attitude the AG was compelled to concede that there was no challenge to the 2G judgement. On that concession, the Supreme Court was persuaded to hear the matter. But as a result, it cut down the number of questions that it would answer to the first five questions, refusing to answer the balance three as they directly concerned the validity of the 2G judgement.

As law minister, I had carefully drafted the questions in close cooperation with the AG and had made a point of including the issue of implications for BITs. Before the formal message of the President reached the Supreme Court, as law minister, I called on Chief Justice S.H. Kapadia to urge him to constitute a constitutional Bench of five judges. Predictably, he was reluctant to spare five judges from the overworked court that he presided over, particularly to bypass the judgement of his colleagues. However, one point persuaded him. I told the Chief Justice that the country would be faced with humongous claims in international arbitrations under BITs. That made the crucial difference and we got the nod for the Reference.

Having extracted the concession that its opinion will not cover spectrum allocation, the Constitutional Bench went on to consider the law applicable to the allocation of natural resources generally.

> To summarize in the context of the present Reference, it needs to be emphasized that this Court cannot conduct a comparative study of the various methods of distribution of natural resources and suggest the most efficacious mode, if there is one universal efficacious method in the first place. It respects the mandate and wisdom of the executive for such matters. The methodology pertaining to disposal of natural resources is clearly an economic policy. It entails intricate economic choices and the Court lacks the necessary expertise to make them. As has been repeatedly said, it cannot, and shall not, be the endeavour of this Court to evaluate the efficacy of auction vis-à-vis other methods of disposal of natural resources. The Court cannot mandate one method to be followed in all facts and circumstances. Therefore, auction, an economic choice of disposal of natural resources, is not a constitutional mandate. We may, however, hasten to add that the Court can test the legality and constitutionality of these methods.
>
> When questioned, the Courts are entitled to analyse the legal validity of different means of distribution and give a constitutional answer as to which methods

are ultra vires and intra vires the provisions of the Constitution. Nevertheless, it cannot and will not compare which policy is fairer than the other, but, if a policy or law is patently unfair to the extent that it falls foul of the fairness requirement of Article 14 of the Constitution, the Court would not hesitate in striking it down.

Finally, market price, in economics, is an index of the value that a market prescribes to a good. However, this valuation is a function of several dynamic variables; it is a science and not a law. Auction is just one of the several price discovery mechanisms. Since, multiple variables are involved in such valuations, auction or any other form of competitive bidding, cannot constitute even an economic mandate, much less a constitutional mandate.

In our opinion, auction, despite being a more preferable method of alienation/allotment of natural resources, cannot be held to be a constitutional requirement or limitation for alienation of all natural resources and therefore, every method other than auction cannot be struck down as ultra vires the constitutional mandate.

Regard being had to the aforesaid precepts, we have opined that auction as a mode cannot be conferred the status of a constitutional principle. Alienation of natural resources is a policy decision, and the means adopted for the same are thus, executive prerogatives. However, when such a policy decision is not backed by a social or welfare purpose, and precious and scarce natural resources are alienated for commercial pursuits of profit-maximizing private entrepreneurs, adoption of means other than those that are competitive and maximize revenue may be arbitrary and face the wrath of Article 14 of the Constitution. Hence, rather than prescribing or proscribing a method, we believe, a judicial scrutiny of methods of disposal of natural resources should depend on the facts and circumstances of each case, in consonance with the principles which we have culled out above. Failing which, the Court, in exercise of power of judicial review, shall term the executive action as arbitrary, unfair, unreasonable and capricious due to its antimony with Article 14 of the Constitution.

In conclusion, our answer to the first set of five questions is that auctions are not the only permissible method for disposal of all natural resources across all sectors and in all circumstances.

As regards the remaining questions, we feel that answer to these questions would have a direct bearing on the mode of alienation of Spectrum and therefore, in light of the statement by the learned Attorney General that the Government is not questioning the correctness of judgment in the 2G case, we respectfully decline to answer these questions. The Presidential Reference is answered accordingly."[96]

---

[96] https://indiankanoon.org/doc/37692759/?type=print

Analysing the Reference

For people who worry about the separation of powers doctrine and often point to an activist Supreme Court's overreach, this opinion is an excellent display of the correct balance. The Court carefully delineated the contours of policymaking domain of the Executive and confined its own jurisdiction to constitutional vires of policy or legislation.

It was argued for PIL (represented by Shanti Bhushan and Prashant Bhushan) that revenue maximization during the sale or alienation of a natural resource for commercial exploitation is the only way of achieving public good since the revenue collected can be channelized to welfare policies and controlling the burgeoning deficit. But the court rejected this view decisively, and held 'we are not persuaded to hold so. Auctions may be the best way of maximizing revenue but revenue maximization may not always be the best way to subserve public good. Common good is the sole guiding factor under Article 39 (b) for distribution of natural resources'.

Considerable effort went into the court answering the Reference but the sum total of the result was that the law stated remains a dead letter. Not for a moment did the Division Bench pause to reconsider the position it had taken. Even if the Advisory Opinion was not binding on the Division Bench it does seem a bit strange and a departure from judicial comity that two judges of the Supreme Court should be impervious to what five other judges, including the Chief Justice, think. Interestingly, besides the Chief Justice, three other judges—Justices Khehar, Dipak Mishra and Ranjan Gogoi—were in line to be chief justices themselves. Technically, the Attorney had conceded that the judgement already pronounced would remain untouched but that applied to the 2007–2008 allocation of spectrum and therefore, strictly speaking, future allocations could well be by methods other than an auction.

The court had indicated that of the eight questions it would answer five and leave the rest as they pertained to the alienation of spectrum and the correctness of the judgement in the 2G case. However, it seems that pressed for time, as the Chief Justice was to retire soon, the court overlooked the fifth question I had very consciously included in the Reference sitting with the AG—the impact of the judgement on investments and BITs. Some weeks later when I brought this to the notice of the new Chief Justice Altamash Kabir he said that he had not yet read the judgement and was handicapped by the fact that his predecessor neither included him on the Constitutional Bench nor kept him in the loop regarding important cases. He suggested that I speak to Justice D.K. Jain, the author of the Reference judgment. I called on Justice Jain who promised to look into the matter but he retired soon after our conversation.

Curiously, while on one hand, the court was willing to countenance methods other than auction, the government was separately examining how to fortify itself against allegations of misfeasance and corruption in allocation of natural resources. For this purpose the Chawla Committee (a high-level committee, headed by former secretary, Finance, Ashok Chawla, favouring auction as the route for disposal of natural resources) was set up and it gave its report in May 2011 on inter alia future telecom licences (to be unified licences and spectrum delinked from licence). Transparent e-auctions for government land in all 81 recommendations were made, of which 69 were accepted by the GoM on corruption headed by the finance minister. The purpose of the exercise was to enhance transparency, effectiveness and sustainability in allocation, pricing and utilization of natural resources through open, transparent and competitive mechanisms. Unfortunately, we got little credit for the effort and for several other legislations that we initiated in response to public sentiment on good governance and transparency.

8
THE FINAL VINDICATION

After the Supreme Court 2G judgement in 2012, there was little chance that a full-fledged trial could be avoided. The court had been monitoring the investigation very closely and had warned against any interference. Even the CBI Special Court was established after the AG gave an undertaking to the court.

Special Judge CBI, O.P. Saini, who pronounced the verdict and was chosen exclusively to handle the 2G trial, began his career as a sub-inspector (SI) in Delhi Police. Hailing from Haryana, Saini sat for the judicial magistrate examination after six years in the police service. Throughout the trial he remained characteristically tough and reluctant to accommodate the accused in matters of bail or permission to travel.

Saini drew up the charge sheet in the 2G matter essentially seeking to charge the accused with the following:

(a) The entry fee for the new pan India UASL in the

year 2008 was kept by DoT as ₹1,658 crore, at which price the CMTS licences were awarded by DoT after auction in the year 2001. These UASLs, issued in 2008 were issued on FCFS basis without any competitive bidding.

(b) A press release was issued by DoT on 24 September 2007 which appeared in the newspapers on 25 September 2007, mentioning that the new applications for UASL will not be accepted by the DoT after 1 October 2007 till further orders. However, applications received only up to 25 September 2007 were considered, which was also against the recommendations of the TRAI that no cap should be placed on the number of Access Service Providers in any service area.

(c) Even the FCFS policy was implemented by the DoT in a manner that resulted in wrongful gain for certain companies. Further, there are allegations that the suspect officials of the DoT had selectively leaked the information to some of the applicants regarding the date of issuance of letter of intent on 10 January 2008. As per conditions laid out in the LoIs, an arbitrary condition that whosoever deposits the fees first, would be the first to get licence, was incorporated. Since some of the applicants, who had this prior information, they were ready with the amount and were able to deposit the fee earlier than others. Thus favour was allegedly shown to some applicants by way of leaking the information about the date of issuance of letter of intent.

(d) Although the FDI limit was increased from 49 per cent to 74 per cent in December 2005, there was no lockin period or restriction imposed on sale of equity or issuance of additional equity. As a result of this, M/s Swan Telecom Pvt. Ltd. (A6), which paid to DoT ₹1,537 crore for UASL of 13 circles, offloaded its 45 per cent equity to M/s Etisalat of UAE for ₹4,200 crore. Similarly, M/s Unitech Wireless (group of eight companies), which paid to DoT ₹1,658 crore for UASL of all 22 circles, offloaded its 60 per cent equity to M/s Telenor of Norway for ₹6,100 crore. These stakes were sold by the said companies even before the roll-out of services by them. The estimated loss to the government by grant of licences to these two companies alone comes to ₹7,105 crore. On pro rata basis, the estimated loss for all 122 UASLs issued in 2008 was more than ₹22,000 crore.
(e) There was quid pro quo between Kalaignar TV, Kanimozhi and the likes of Shahid Balwa in the ₹200 crores being given on loan to the TV Channel.
(f) Clause 8 of the Regulations was violated to accommodate applicants who were not qualified to apply under the ToR.

At Patiala House Court, Judge Saini pronounced his 2G verdict in just a single line: 'The prosecution has miserably failed to prove its case, and all accused are acquitted.'

In the 1,552 pages packed with detailed analysis of departmental files and scrutiny of testimony of several key civil servants, the trial court found no evidence to substantiate the charges against the accused.

One clever device employed by the adversaries of the UPA was to attempt to drive a wedge between the minister under siege and the prime minister. The court was not impressed as highlighted in the following judgement of O.P. Saini[97]:

> Letter to Prime Minister: Misrepresentation of Facts and Misleading the Prime Minister:
>
> Para 628.
>
> It is the case of the prosecution that A. Raja decided the cutoff date of 25.09.2007 vide note dated 02.11.2007, Ex PW 36/B8 (D7) and on deciding this date, he wrote a letter dated 02.11.2007, Ex PW 7/A, to the Hon'ble Prime Minister misrepresenting the facts and fraudulently justifying the cutoff date of 25.09.2007.[98]
>
> Para 643.
>
> However, if a Minister writes directly to the Hon'ble Prime Minister, that does not by itself mean that Prime Minister would be misled. Furthermore, there is no material on record indicating that a Minister cannot directly write to the Prime Minister.[99]

---

[97]https://www.hindustantimes.com/static/ht2017/12/CBI%20Vs.%20A.%20Raja%20and%20others.pdf
[98]CBI v. A. Raja and others, Case RC No. 45 (A) 2009, CBI, ACB, New Delhi, para 628; http://www.thehindu.com/news/national/tamil-nadu/article22122355.ece/BINARY/2GCBIVsRaja
[99]CBI v. A. Raja and others, Case RC No. 45 (A) 2009, CBI, ACB, New Delhi, para 643; http://www.thehindu.com/news/national/tamil-nadu/article22122355.ece/BINARY/2GCBIVsRaja

Para 644.

Accordingly, this is no ground to say that Hon'ble Prime Minister was misled by Sh. A. Raja, as he wrote the letters on his own without getting the same processed in the DoT files. In the end, I do not find any merit in submission of the prosecution that the Hon'ble Prime Minister was either misled by Sh. A. Raja or that the facts were misrepresented to him. The arguments have been taken up by the prosecution just to prejudice the mind of the Court by invoking the high name and authority of Hon'ble Prime Minister of the country.[100]

The change in cut-off date was perhaps the most intriguing and persuasive allegation with a 'no smoke without fire' touch to it. But this charge too failed miserably:

Para 446.

In view of the above facts, there is no material on record to show that the note regarding cutoff date was initiated by Sh. A.K. Srivastava at the initiative of Sh. A. Raja, conveyed through Sh. R.K. Chandolia to benefit the accused companies.[101]

---

[100] CBI v. A. Raja and others, Case RC No. 45 (A) 2009, CBI, ACB, New Delhi, para 644; http://www.thehindu.com/news/national/tamil-nadu/article22122355.ece/BINARY/2GCBIVsRaja

[101] CBI v. A. Raja and others, Case RC No. 45 (A) 2009, CBI, ACB, New Delhi, para 446; http://www.thehindu.com/news/national/tamil-nadu/article22122355.ece/BINARY/2GCBIVsRaja

Para 447.

The central issue here is not whether the note proposing cutoff date is right or not, but whether it is result of criminal conspiracy being executed by Sh. A. Raja and Sh. R.K. Chandolia through the innocent agency of Sh. A.K. Srivastava. The answer to the question is an emphatic 'No', as the note was not put up by Sh. A.K. Srivastava either under pressure or on the asking of Sh. R.K. Chandolia, but on his own initiative after discussion within the department. It is thus clear from the evidence, that putting up of note, Ex PW 36/E1, regarding cutoff date of 10.10.2007, on account of receipt of large number of applications for UASL or its curtailing by Sh. A. Raja to 01.10.2007, was not the result of any conspiracy, but was an administrative step taken up by the officers of DoT in view of receipt of large number of applications, but was later on disowned by them when the issue became controversial. The facts examined thus far do not reveal any conspiracy. Sh. A. Raja had approved the cutoff date by citing three reasons, that is, pendency of large number of applications, to discourage speculative players and time of one month from the receipt of TRAI Recommendations. These are good reasons, if seen in the light of the note, Ex PW 36/E1, dated 24.09.2007 recorded by Sh. A.K. Srivastava. The date of 10.10.2007 or 01.10.2007 would not have made any difference to Unitech group of companies as their applications had already been filed on 24.09.2007. STPL had already applied as early as 02.03.2007.

Pg. 486:

> Conclusion: The conclusion from the above detailed discussion is that there is absolutely no evidence on record that the very concept of cutoff date or the cutoff dates of 01.10.2007 or 25.09.2007 are the result of any conspiracy by the conspiring public servants, that is, Sh. A. Raja and Sh. R.K. Chandolia with Shahid Balwa, Vinod Goenka and Sanjay Chandra. Entire submission of the prosecution is without merit.[102]

The policy of First Come First Serve (FCFS) was obviously inevitable once auction was ruled out. Here the allegation was that Raja modified that to First Comply First Serve. The court's finding is clear:

Para 770

> The end result is that prosecution has failed to prove that there was any policy of first come first serve as alleged by it and if there was such a policy that it was being followed by DoT in the manner alleged by the prosecution. Thus, the policy of first come first serve remained only an abstract proposition born out of necessity, the details of which were not clear to anyone. It was also violated first by the Minister, who had laid it down. This violation also added to ambiguity in the policy leading to many interpretations. The

---

[102]CBI v. A. Raja and others, Case RC No. 45 (A) 2009, CBI, ACB, New Delhi at para 486; http://www.thehindu.com/news/national/tamil-nadu/article22122355.ece/BINARY/2GCBIVsRaja

prosecution case can fail on this ground alone.[103]

The issue of Entry Fee, the core of the case for Presumptive Loss, was examined by the court from two points of view: (1) Whether the TRAI suggested that there was no justification for fresh market discovery of licence fee; (2) whether the Ministry of Finance insisted that auction was necessary. In both the trials, the court found Raja and others accused not to be in error, let alone any criminal culpability.

Objection of Ministry of Finance relating to Revision of Entry Fee:[104]

> Para 1623. It may be noted that Finance Ministry did not revert to the DoT, what to talk of persisting with its objection of nonrevision of entry fee. This was despite the fact that reply of DoT, Ex PW 78/C, was brought to the notice of the then Finance Minister Sh. P. Chidambaram on 30.11.2007 and his specifically asking the Ministry vide his note Ex PW 78/DB to examine the two issues carefully, that is, the issuance of UAS licence at an entry fee discovered in 2001 and TRAI not recommending revision of entry fee in its Recommendations dated 28.08.2007. From the above discussion, it is clear that Ministry of Finance was not very enthusiastic about its objections regarding pricing of initial spectrum/revision of entry fee. Moreover, Finance Secretary admitted that after receipt of reply of DoT they did not pursue the

---

[103]CBI v. A. Raja and others, Case RC No. 45 (A) 2009, CBI, ACB, New Delhi, para 770; http://www.thehindu.com/news/national/tamil-nadu/article22122355.ece/BINARY/2GCBIVsRaja
[104]Ibid.

objections seriously. If the Finance Ministry had been serious and Sh. A. Raja was not heeding to its query for revision of entry fee, the matter must have been reported to the Cabinet Secretariat or PMO. However, there is no material on record in this regard. Accordingly, the argument of the prosecution that the Ministry of Finance and Member (F) had objected to the nonrevision of entry fee and that was not heeded by Sh. A. Raja is without merit. The deposition of Sh. D. Subba Rao is of no use to the prosecution as he took an objective view of things and as such there is nothing incriminating therein.

Para 1624

It is thus clear that Ministry of Finance did not pursue the matter further and the note of Member (F) was absolutely uncalled for. Thus, the letter of Finance Secretary does not advance the case of prosecution.

On whether the Issue of Entry Fee is considered by DoT.[105]

Para 1625

As already noted above, it has been the case of the prosecution that the issue of entry fee was never considered in the DoT and the finance branch had no occasion to offer its views on it. Let me consider the same in detail to clarify the position.[106]

---

[105]CBI v. A. Raja and others, Case RC No. 45 (A) 2009, CBI, ACB, New Delhi, para 1624; http://www.thehindu.com/news/national/tamil-nadu/article22122355.ece/BINARY/2GCBIVsRaja
[106]CBI v. A. Raja and others, Case RC No. 45 (A) 2009, CBI, ACB, New

The Conclusion drawn from the judgement was:[107]

> In the end, there is no merit in the submission of the prosecution that the FCFS policy was manipulated by the accused to the benefit of two accused companies and that Hon'ble Prime Minister was misled on this point. There is also no merit in the submission that the change of policy was manipulated by the accused and the accused beneficiary companies had prior knowledge of it. The entire prosecution case on this point is without merit.
>
> That most of the mess in DoT, in the matter of processing of applications for UAS licences, and grant of licences was created by the officers. It is the result of their lack of sense of responsibility and clarity about the way official business is to be conducted. Not only this, most of the officers have exhibited fickle mindedness and timidity by disowning the written official record. They even disowned the record prepared by them and tried to shift the blame to others by making oral statements contrary to official record. Official record prepared by a public servant in the ordinary course of business is deemed to be correct and truthful and is considered sacrosanct. It is more so in the case of officers as senior as Joint Secretary, Special Secretary and the Secretary to Government of India, because they

---

Delhi, para 1625; http://www.thehindu.com/news/national/tamil-nadu/article22122355.ece/BINARY/2GCBIVsRaja

[107] https://www.scribd.com/document/367660716/2G-spectrum-allocation-scam-verdict-Full-text#from_embed

constitute the core of governance in the country. Their actions become precedent for the future. However, all officers of such superior ranks, endeavoured hard to disown their own notes and blame others, that is, the accused, for everything done by them. This is not acceptable in the face of the official record. These witnesses kept wavering and were not committed to any particular stance. Their evidence turned out to be unworthy of reliance. The conduct of the above officials deserves strong disapproval.

The lack of clarity in the policies as well as guidelines also added to the confusion. The guidelines have been framed in such technical language that meanings of many terms are not clear even to DoT officers. When the officers of the department themselves do not understand the departmental guidelines and their glossary, how can they blame companies/others for violation of the same. The worst thing is that despite knowing that the meaning of a particular term was ambiguous and may lead to problems, no steps were taken to rectify the situation. This continued year after year. For example, in the instant case, large part of the controversy relates to interpretation of Clause 8, dealing with substantial equity. The terms used in this clause include 'Associate', 'Promoter', 'Stake', etc. No one in the DoT knows their meanings, despite the fact that the guidelines were framed by the DoT itself. The interpretation of these words is haunting the DoT since these words were first used, but no steps were taken to assign them a specific meaning. In such circumstances, DoT officers themselves are responsible for the entire mess.

There is nothing to show that A. Raja was mother lode of conspiracy in the instant case. There is also no evidence of his no holds barred immersion in any wrongdoing, conspiracy or corruption.

There is no evidence on the record produced before the court indicating any criminality in the acts allegedly committed by the accused persons relating to fixation of cut-off date, manipulation of FCFS policy, allocation of spectrum to dual technology applicants, ignoring ineligibility of STPL and Unitech group companies, non-revision of entry fee and transfer of ₹200 crore to Kalaignar TV (P) Limited as illegal gratification. The charge sheet of the instant case is based mainly on misreading, selective reading, nonreading and out of context reading of the official record. Further, it is based on some oral statements made by the witnesses during investigation, which the witnesses have not owned up in the witness box. Lastly, if statements were made orally by the witnesses, the same were contrary to the official record and thus, not acceptable in law.

I may add that many facts recorded in the charge sheet are factually incorrect, like Finance Secretary strongly recommending revision of entry fee, deletion of a clause of draft LoI by Sh. A. Raja, Recommendations of TRAI for revision of entry fee, etc.

The end result of the above discussion is that, I have absolutely no hesitation in holding that the prosecution has miserably failed to prove any charge against any of the accused, made in its well-choreographed charge sheet.

Judge Saini disposed of two other matters linked to the main 2G prosecution. He may well have thought that this would be the last he had see of the 2G matter, having also discharged the Maran brothers in the related Aircel-Maxis case. But that was not to be. Like bad losers, CBI and ED continued to pursue Karti Chidambaram and the former finance minister. Despite the discharge granted by Judge Saini to the Maran brothers, the agencies still believe they can make out a case of corruption in the FIPB clearances. So 2G was back in Judge Saini's court when Karti Chidambaram's bail matter came up. Karti was unfairly arrested by the CBI at Chennai Airport upon return from abroad having obtained permission from the Madras High Court. He was merely required to answer any summons for appearance upon his return. Yet he was arrested and frisked away to the CBI headquarters in New Delhi. While he was in judicial custody and hoping to get bail there was an apprehension that the ED would step in with PMLA as soon as bail was granted. Thus they were back in Judge Saini's court for another drama. The court kept asking if there was any intention to execute a fresh arrest and the answer was repeatedly, 'not as of this moment'. Ultimately, bail was granted and the larger issue travelled to the Supreme Court after a brief hearing in the High Court.

Epilogue
CURTAIN CALL

Even as the heat and dust of 2G settles down, and truth—endorsed by trial—begins to displace the pall of falsehood and fake perceptions, the print lines and airwaves are getting crowded with talk of Vyapam, Rafael, Gujarat Land scam, Judge Loya's death et al. There is no word on the much awaited Lokpal for which a relentless public agitation was engineered by the BJP and Anna Hazare towards the end of the UPA II tenure. Governments formed by parties who shouted themselves hoarse on combating corruption have done precious little to even take to logical conclusion the steps that we had taken before demitting office. What we do get to hear repeatedly are self-serving pronouncements of integrity. A lie spoken often enough is said to become true, we are told. Fortunately, the Congress had introduced the far-reaching Right to Information that has made an irreversible impact on our ability to access information in the State's possession, information that affects our lives. Corruption of course is not limited to financial matters and indeed moral corruption is an even greater threat to humanity. The repeated incidents of lynching for communal reasons and the growing incidence of heinous and violent crime against women and the girl child have blotted the face of our society. The public response from the top leadership is an inexplicable silence or belated, insipid regret. Meanwhile, lesser leaders continue making divisive and humiliating remarks about citizens who disagree with them. As the true patriots feel a sense of outrage, the nation holds its breath, not knowing what tomorrow holds for us.

As the noise and outrage grow across the country, the matter of the 2G judgement has retreated into routine oblivion and will rest there for a while till the High Court can find time to address it by way of an appeal. The fact is that there cannot be another trial now and it is only if the High Court finds that on the evidence produced before the Trial Court another view was possible, is compelling. In the language of the law, the decision of the Trial Court will have to be 'perverse' to be overturned by the Appellate Court. Interestingly, Justice G.S. Singhvi made a brief comment to the media about the Trial Court judgement emphasizing that the Supreme Court decision was on different grounds from the ones on which this was based. One has heard no apology from Arvind Kejriwal unlike the spate of apologies he has given to settle dozens of defamation cases.

The entire saga of 2G reads like 'kabhie khushi kabhie gham' (a roller coaster of both joy and grief). That might have had a particular impact on the lives of many people who were accused as indeed on the lives of many others who were swept away by the public opinion tsunami, but ultimately there is no takeaway in terms of better understanding of the working of the government. Both in acquitting civil servants like Behura and in expressing displeasure about several others, including those who gave oral testimony, Judge Saini did not use the opportunity to discover how the bureaucratic structure of the executive works. The inability to understand the structure and passing judgement with overarching consequences can cause irreparable damage to the legendary steel frame of the government. People familiar with the system will tell you that it is essentially built on trust between the political executive and the bureaucratic structure. Any step that undermines the basis of this trust will inevitably undermine the quality of governance just as respect by the executive and the people is fundamental to the final arbiter, the judicial system and its independence.

The end result is that no one seems any wiser. The people who suffered incarceration and severe damage to their reputations will never be compensated adequately and it is unlikely that the entire episode will have a deterrent impact on future transactions, and the public will just let the matter be forgotten as fresh supposed scandals are exposed by the media. Will the experience teach us something we did not know about the government? The pursuit of good governance and integrity in public life is indeed high national priority but in dogged commitment to it one cannot entirely overlook the equally important, if not more, equal justice according to law. The distinction between rule of law and rule of men (and women) has therefore been a sentinel point in liberal democracies. No matter how significant the benefit to be gained for the people at large it does not justify tinkering with the system. The words of Thomas More to his son-in-law are edifying:

> William Roper: So, now you give the Devil the benefit of law!
>
> Sir Thomas More: Yes! What would you do? Cut a great road through the law to get after the Devil?
>
> William Roper: Yes, I'd cut down every law in England to do that!
>
> Sir Thomas More: Oh? And when the last law was down, and the Devil turned 'round on you, where would you hide, Roper, the laws all being flat? This country is planted thick with laws, from coast to coast, Man's laws, not God's! And if you cut them down, and you're just the man to do it, do you really think you could stand upright in the winds that would blow then? Yes, I'd give the Devil benefit of law, for my own safety's sake!

In competitive politics, parties often lose sight of the larger picture and take positions that come back to haunt them another day. Howls of 'hypocrisy' go up each time much to the chagrin and surprise of the incumbent. Recent washouts in Parliament are living examples of this. At some point political parties will have to take a pause and reflect on the damage being done by their respective intransigence. Politics must get back to creative competitiveness and give the people of India substantive choice. However, the pragmatic reality is the 2019 electoral contest that promises to be like never before. There will of course be several critical issues, the failures of the present dispensation not being the least, but considerable coordination between like-minded opposition parties will be the tipping point. Congress under a youthful, charismatic Rahul Gandhi will come equipped with a makeover. But even as the ruling party pulls out all stops to damage the Congress, our party will have to shake itself free of the slings and arrows of contrived misfortune inflicted upon us by the adversaries. The Phoenix must indeed rise from the ashes.

The dramatis personae of the 2G saga have quietly exited from the stage and returned to their normal existence. Only Raja chose to place his story and his point of view before the people in an autobiographical account, *2G Saga Unfolds*. The former minister's tone suggests that adversity has mellowed him somewhat but understandably, he has not forgiven Vinod Rai. 'I sincerely hope that I have clearly and factually demonstrated Mr Rai in his unscrupulous quest for self-aggrandisement, betrayed the community confidence and committed condemnable sacrilege'. To borrow a phrase from Winston Churchill with an apology: 'Never in the history of human endeavour would so much have been damaged and destroyed for so many by so few.'

Some of the physical debris has been cleared by time, some moved in salvage operations by the UPA and its successor NDA, but dust will not disappear till the last gavel of law concludes the electoral contest.

ACKNOWLEDGEMENTS

There was enormous material to go through for this book, including legal material. My colleagues in my law chambers all helped in one way or the other, directly and indirectly. I spent time talking to a dear friend Navaid Khan and his friends from the world of telephony and gathered interesting tales of the sunshine sector, not all of which they would want me to reproduce.

My colleagues in public life who were more directly involved in looking at the entire 2G story knowingly and unknowingly helped me sift the wheat from chaff of telephony.

ABBREVIATIONS

AG Attorney General
AGR Adjusted Gross Revenue
ARPU Average Revenue Per User

BICP Bureau of Industrial Cost & Pricing
BITs Bilateral Investment Treaties
BSNL Bharat Sanchar Nigam Ltd.
BTS Basic Telephone Service
CAG Comptroller and Auditor General of India
CDMA Code Division Multiple Access
C-DOT Centre for Development of Telematics
COAI Cellular Operators Association of India
CMSP Cellular Mobile Service Providers
CMTS Cellular Mobile Telephone Services
CPIL Centre for Public Interest Litigation
CVC Central Vigilance Commission

DoT Department of Telecommunications
ED Enforcement Directorate

EMF Electromagnetic Field

FBG Financial Bank Guarantee
FCFS First Come first serve
FDI Foreign Direct Investment
FIPB Foreign Investment Promotion Board

GoM Group of Ministers
GoT Group on Telecom
GoT-IT Group on Telecom and IT
GSM Global Systems for Mobile Communications

IAC India Against Corruption
ICJ International Court of Justice
ICT Information and Communications Technology
IOA Indian Olympic Association
ITU International Telecommunication Union

LoI Letters of Intent

MoC&IT Ministry of Communications and Information Technology
MTNL Mahanagar Telephone Nigam Ltd.

NFAP National Frequency Allocation Plan
NTP New Telecom Policy

PBG Performance Bank Guarantee
PCA Prevention of Corruption Act
PMO Prime Minister's Office

PSU Public Sector Undertakings

SDCA Short Distance Charging Area

TDSAT Telecom Disputes Settlement and Appellate Tribunal
ToR Term of References
TRAI Telecom Regulatory Authority of India

UASL Unified Access Service Licencse

WLL Wireless in Local Loop
WPC Wireless Planning and Co-ordination Wing in DoT

INDEX

Aadhaar, 4, 9
Aadhaar-linked direct benefits, 2
Aam Admi Party (AAP), 6, 8
Access Service Licences, 86
Access Service Providers, 102, 146
Active brain, *See* Kanimozhi, M.K.
Administrative allocation model, 25
Adonis Projects, 40, 84
Agarwal, Rajiv, 47
Aircel, 30, 36, 127, 129–30, 157
Aircel-Maxis case, 127, 157
Airtel, 25, 30, 34, 36, 76, 79
Alienation mode of Spectrum, 141
Allianz Infratech, 42, 114
Allocation of Spectrum, 14–17, 66

administrative approach, 14
auction, 15
fair and transparent allocation, 16
first-come-first-serve (FCFS) basis, 15
lottery method, 15
spectrum management, 15
Arbitrary changes, 57
Aska Projects, 40
Association of Unified Telecom Service Providers of India (AUSPI), 36
Astro All Asia, 127–8, 130
Astro Holdings, 128
Auction, action for conduct of, 137
Average revenue per user (ARPU), 22
Azare Properties, 83
Azka Projects, 83

Azure Properties, 40

Balwa, Asif, 47
Balwa, Shahid, 40, 44, 47, 49, 128, 147, 151
Basic Service Licensees, 78
Basic Service Operators (BSOs), 52, 98
Bedi, Kiran, 8
Behura, Siddharth, 44–5, 104
Bharat Sanchar Nigam Ltd. (BSNL), 23, 30, 97
Bharti Airtel, 76, 79, See also Airtel
Bhushan, Prashant, 113, 142
Bhushan, Shanti, 142
Bilateral Investment Treaties (BITs), 127, 138–9, 143
BTS licences, 19, 24, 98
Bureau of Industrial Cost & Pricing (BICP), 92

CAG Report, 3, 33, 50, 63–4, 89–109, 133
  assessment of the financial impact, 105
  critical issuance of the UAS licences, 100
  loss due to Migration Package, 94
  meaning and role of spectrum, 107
  performance audit report, 90
  private sector participation, 91
  result of privatization, 93
  role of TRAI, 106
  UAS licence regime, 98
  unified licensing, 99–100
Cancellation of the licences, 114, 126–7
CDMA-based systems, 27–8
CDMS spectrum, 31
Cellular Mobile Service Operators (CMSOs), 52
Cellular mobile service providers (CMSP), 22
Cellular Mobile Telephone Services (CMTS), 17, 19, 23–5, 69–70, 77, 79, 91, 96–7, 103, 146
Cellular Operators Association of India (COAI), 21, 92
Central Bureau of Investigation (CBI), 2, 37, 43–6, 49, 90, 105, 124, 127–30, 145, 148–53, 157
Central Vigilance Commission (CVC), 90
Centre for Development of Telematics (C-DOT), 66
Centre for Public Interest Litigation (CPIL), 29, 113
Chacko, P.C., 90
Chandolia, R.K., 45, 149–51
Chandra, Sanjay, 45, 151

Changes in policy to favour cellular operators, 97
Chawla Committee, 143
Chawla, Ashok, 143
Chidambaram, Karti, 130, 157
Chidambaram, P., 7, 111, 127, 130–1, 134, 152
Churchill, Winston, 163
Code Division Multiple Access (CDMA), 24, 27–8, 30, 36, 103, 108–9, 137
Competition Appellate Tribunal, 112
Comprehensive spectrum reform, 86
Comptroller and Auditor General (CAG) report, *See* CAG Report
Concessions, 92
Constitutional Bench, 3, 15, 132–3, 138–9, 143
Criminal proceedings, 8, 132
CVC, 37, 90
CWG controversy, 4

Damage control, 39
Datacom (Videocon), 41, 63
DB Realty, 40, 44, 47, 49, 128
Decisive mandate, 9–10
Demonetization, 9, 35
Dikshit, Sheila, 4–5
Doshi, Gautam, 46
Dual spectrum, 71, 102
Dual technology, 33, 36, 63, 72, 103, 156
Dual Technology licences, 28–9, 59–62, 137
Duttu, H.L., 46

e-auctions, 144
Election Commission, 96
Electoral contest (2019), 163
Enforcement Directorate (ED), 129–30, 157
Entry Fee, 94, 152–3
  realistic reassessment of, 121
Essar Group, 48–9
Essar Tele Holding, 49
Etisalat, 40, 63, 128, 147

Fairness and transparency, principles of, 78–9, 81, 84
Financial Bank Guarantee (FBG), 58
First Come First Serve (FCFS), 15, 25, 27, 29, 33, 37, 39, 43, 58, 65, 70–2, 74, 76, 78–9, 82, 85, 87, 90, 100, 103, 113, 116–17, 121, 124, 146, 15, 154, 156

First Comply First Serve, 39, 151
Foreign Direct Investment (FDI), 4, 26, 61, 81, 131, 135, 147

Foreign Investment Promotion
  Board (FIPB), 81, 101,
  130-1, 157
Full Telecom Commission, 86,
  98, 117

Gandhi, Rahul, 163
Ganguly, A.K., 124
Ganguly, Asok Kumar, 114
Goenka, Vinod, 47, 49, 128,
  151
Gogoi, Ranjan, 142
Gokak, A.V., 92
Goods and Services Tax
  (GST), 4, 9
Government of India
  (Transaction of Business)
  Rules, 115
Group on Telecom (GoT), 22,
  95, 98
Grover, Anand, 125
Gujarat Land scam, 159

Hazare, Anna, 5, 159
Hudson Properties, 40, 84

Idea Cellular, 30, 36, 42, 75-6,
  84
Income Tax Act, 124
India Against Corruption
  (IAC), 5
Indian Olympic Association
  (IOA), 5

Indian Penal Code, 48
Ineligible Applicants, 59
Internal Telecom Commission,
  86
International Court of Justice
  (ICJ), 128
International Radio
  Regulations, 69, 108
International
  Telecommunication, 13
Irregularities, 32
ITU guidelines, 68

Jain, D.K., 143
Jayalalithaa, J., 63
Joint Parliamentary Committee
  (JPC), 3, 88-91, 109,
  133
Judge Loya's death, 159
Judicial review, 120, 122, 132,
  136, 141

Kabir, Altamash, 125, 143
Kalmadi, Suresh, 4-5
Kanimozhi, M.K., 43-4
Kapadia, S.H., 125, 135, 138
Karunanidhi, M., 43, 63
Kejriwal, Arvind, 5-8, 160
Khaitan, I.P., 48
Khaitan, Kiran, 48
Koshika Telecom, 18
Krishnan, T. Ananda, 130
Kumar, Meira, 89

Kumar, Sharath, 48

Lalwani Committee, 76
Level-playing field, 56
Licence fee, outstanding dues, 94
Lokpal, 5, 159
Loop Telecom, 41, 48–9, 63, 114
Love-jihad, 9
Lynching for communal reasons, 160

Mahanagar Telephone Nigam Ltd. (MTNL), 23, 30, 66, 97
Mahatma Gandhi National Rural Employment Guarantee Act (MGNREGA), 2
Malaysia's Astro Asia Networks Ltd. (AAANL), 127
Maran, Dayanidhi, 127, 129–30
Maran, Kalanithi, 129
Maxis Communications, 127, 130
Migration Package, 94–7, 100
Ministry of Law and Justice, 55
Mishra, Dipak, 142
Mishra, Mahabal, 7
Mishra, Nripendra, 37

Mobile Telephony Industry, 11, 124–7
Money-laundering case, 129
Monopoly, 16, 18
Morani, Karim, 48
Mukherjee, Pranab, 134

Nahan Properties, 40
Nair, Hari, 46
National Frequency Allocation Plan (NFAP), 69, 108
National Rural Health Mission (NRHM), 2
Natural justice, principles of, 76
Natural resources
  alienation of, 140
  distribution of, 139, 142
  pricing and utilization of, 144
New Economic Policy (1991), 106
New Telecom Policy (1994), 67, 91, 93
New Telecom Policy (1999), 22–3, 26, 33, 38, 66–8, 70, 74, 87, 95–8, 100, 106, 119–20
'No smoke without fire', 149
Non-performing asset (NPA), 32

Occupy Wall Street, 6

Once-bitten-twice-shy
approach, 31
One Man Committee (OMC),
65–6, 87–8

Patil, Pratibha Devisingh, 135
Patil, Shivraj, 3, 65, 133
Pawar, Sharad, 5
Penalties, 122, 132
Performance Bank Guarantee
(PBG), 58
Pipara, Surendra, 46–7
Polarization politics of
communities, 9
Policy shift, 96
Presumptive loss, 3, 40, 62–3,
105, 152
Prevention of Corruption Act
(PCA), 40, 43–9
Prevention of Money
Laundering Act, 124
Public Accounts Committee,
63
Public Interest Litigation (PIL),
125, 131, 142
Public servants, culpability of,
66
Publicized auction, 121

Radio Communications Act,
87
Radio Spectrum, 12–14
Radiolinja, 16

Rafael, 159
Rai, Vinod, 50, 63, 163
Raja, A., 3, 38–40, 43–5, 63–4,
71, 87, 90, 111, 114, 124,
148–53, 156, 163
Regulators, 12, 14–15
Reliance Anil Dhirubhai
Ambani Group, 46
Reliance Communications, 30
Reliance Telecom, 36, 49
Resettlement and Welfare
Schemes of the Ministry
of Defence, 123
Revenue maximization, 16, 142
Revenue sharing, 22, 96
Right to Education (RTE), 2
Romeo squads, 9
Ruia, Anshuman, 48
Ruia, Ravi, 48

S Tel, 30, 42, 63, 105, 114
SAHEL companies, 128
Saini, O.P., 2, 110, 130, 145,
148
Saraf, Vikas, 48
Satpal Maharaj, 7
Separation of powers, 142
Shivraj Patil Committee
Report, 133
Short Distance Charging Area
(SDCA) mobility, 98
Shourie, Arun, 76
ShyaniTelelink, 41

Sibal, Kapil, 87, 110, 134
Sikri, A.K., 133
Singh, A.K., 46
Singh, Manmohan, 6, 37, 63, 71
Singh, V.K., 8
Singhvi, G.S., 3, 112, 114, 125, 132–3, 160
Sistema Shyam, 34, 41, 114
Sivasankaran (Siva), 127, 129
South Asia Entertainment Holdings Ltd., 130
Spectrum allocation, 7, 13, 15, 22, 24–5, 27, 29, 31, 52, 86, 102, 118, 139
Spectrum auction, 17–22
  basic telephone service (BTS), 19
  criteria, 17
  licence fee, 18
  pricing and allocation, 21
  technical and financial evaluation, 17
  TRAI's recommendations, 20
  usage charges, 19–20
Spectrum planning, 13
Spectrum vacation, 54
Spice Communications, 42
Standard option for pricing, 120
State monopoly, 106
Stockholm syndrome, 9
Subscriber-based criterion, 25

Sun Direct TV (P) Ltd. (SDTPL), 127, 130
Sundararajan, Aruna, 33
Supreme Court Legal Services Committee, 123
Swamy, Subramanian, 113, 118, 130
Swan Telecom, 39, 42, 47, 49, 114, 147

Tata Docomo, 30
Tata Teleservices, 30, 42, 75, 79, 114
TDMA-based system, 27–9
Telecom Commission, 53–4, 56, 69, 80, 86, 93, 97–9, 102, 115
Telecom Disputes Settlement and Appellate Tribunal (TDSAT), 21, 95, 98, 100, 107, 128
Telecom Regulatory Authority of India (TRAI), 20–3, 26, 28, 30, 36–8, 52–4, 56–7, 60, 62, 64, 68, 70, 86–7, 92, 96–100, 102–3, 105–7, 113–15, 118–22, 137, 146, 150, 152, 156
Telecommunications Commission, 66
Telenor, 39, 128, 147
Term of References (ToRs), 54
3G Spectrum auction, 29–31

TRAI Act, 20-1, 92, 107
TRAI recommendations,
    30, 37-8, 54, 68, 71, 74,
    102-3, 114, 123, 137
Transparency, 51, 59, 68, 76-7,
    82, 85, 116, 144
Trump, Donald, 9
2G judgement, implications of
    the, 127
2G Saga Unfolds, 163
2G trial court, 32

UAS Licence, 26-9, 38, 52, 54,
    58-60, 65, 69-73, 76-80,
    82-3, 99-100, 103-4,
    113-14, 120, 123, 145-7,
    150, 154
  implementation of, 52
  policy of, 51
  revised guidelines, 26-7
  TRAI recommendations, 26
  unified licensees, 26

upfront charge, 27
Unitech Builders, 39-41, 83
Unitech Wireless, 45, 49, 114,
    147
Unlicenced frequency bands,
    14
UPA-I, 1-2, 4
UPA-II, 1-2

Vahanvati, Ghulam, 39, 112
Vajpayee, Atal Bihari, 2, 22
Venugopal, V., 132
Videsh Sanchar Nigam Ltd.
    (VSNL), 66
Vodafone, 30, 34, 36
Volga Properties, 40, 83
Vyapam, 159

Wireless Planning and Co-
    ordination (WPC), 69

Zero loss, See Sibal, Kapil